FOUNDATIONS OF DIATONIC THEORY

A Mathematically Based Approach to Music Fundamentals

Timothy A. Johnson
Ithaca College

Key College Publishing
Innovators in Higher Education

www.keycollege.com

CONTENTS

PREFACE

"Introductory pedagogical practice is guided less by current research in music theory than by the speculative theory of past centuries. In the twentieth century, these concepts . . . have run deep grooves in the pavement, grooves which much current introductory pedagogy retraces."

—Richard Cohn
Music Theory's New Pedagogability, 1998

Although most of the material in this text was developed before the publication of Richard Cohn's article, this book, in part, begins to answer Cohn's call for a stronger link between introductory pedagogy and recent scholarship in music theory. By relating concepts in diatonic set theory directly to the study of music fundamentals through pedagogical exercises and instruction, this book exposes the undergraduate student to results of the most recent scholarship. In this way *Foundations of Diatonic Theory* attempts to steer clear of the deep grooves of the historical legacy without leaving the road altogether.

The pedagogical material in this text was originally designed for use as a supplement in traditional Theory I courses, but in its current form it is equally appropriate for courses in the fundamentals of music (for music majors or for non-majors) and for stand-alone courses involving the integration of mathematics and music. It is also well-suited for self-study because all of the exercises provided are solved and discussed directly in the text. This feature allows students to check their work as they make their way through the text. Solutions to the exercises also can be reviewed quickly by instructors, leaving plenty of class time for other topics.

The study of diatonicism by means of set theoretical tools has been a relatively new focus in the literature, and interest in the field of diatonic set theory has increased rapidly over the past decade or so. In addition to its primary audience of undergraduates, this text also provides an accessible point of entry into the field for scholars, professional musicians, or graduate students, who should find both the ideas and the pedagogical implications attractive.

$\mathcal{T}$O THE INSTRUCTOR

Course Planning

Foundations of Diatonic Theory augments a variety of classroom situations: a course in music fundamentals, either for majors or non-majors; the review of fundamentals in any course in the core-theory sequence; a course in music and mathematics; or an advanced course in diatonic set theory. Music fundamentals courses, covering the building blocks of music throughout a semester, can introduce the material in this book concurrently with the applicable concepts. On the other hand, core-theory courses may review the related fundamentals material faster than the time needed to complete this text, so that much of this text may be completed individually by students as the class moves on to other topics. The table that follows shows the minimum music theory concepts necessary to complete each chapter; other concepts are introduced as needed within the text. A more detailed discussion of the music theory concepts associated with each part of this book is provided in the *Instructor Resources*.

MINIMUM MUSIC THEORY CONCEPTS NECESSARY TO COMPLETE EACH CHAPTER	
Chapter	**Concepts**
1. Spatial Relations and Musical Structures	major and minor scales whole and half steps
2. Interval Patterns and Musical Structures	intervals key signatures circle of fifths
3. Triads and Seventh Chords and Their Structures	triads and seventh chords

In core-theory or music fundamentals classroom situations, little extra time (perhaps five to ten minutes per class session) is required to introduce and specifically discuss this material. The organization fits well with the usual introductory topics in music theory, and solutions and full explanations are provided for all of the exercises in the text. The reading and exercises contained in this book can be discussed in short segments at the beginning or end of class, without disrupting the usual flow of fundamentals material—and in many cases can take the place of more tedious review exercises.

A course devoted to mathematics and music might use this text for one or more units of the course. Due to the orientation of this text, relating concepts in diatonic set theory directly to music fundamentals, mathematically minded students will find a good introduction to basic musical concepts. The materials at the end of the text (For Further Study, Notes, and Sources Cited) reference some of the sources that treat this material in a more mathematically rigorous way—involving intricate formulas, proofs, and other advanced mathematical constructs—for students and teachers who wish to explore these and other related ideas in more depth.

This text also provides a fitting introduction for an advanced course in diatonic set theory. By studying this textbook first, students can approach some of the key principles of diatonic set theory in a familiar context—music fundamentals—before encountering the more formal orientation of the literature. This textbook might occupy only the first few weeks of class time in an advanced course in diatonic theory, but it would provide a firm foundation for the entire course.

Teaching the Course

In teaching these materials in introductory music theory and fundamentals classes, I have found that beginning students have reacted positively to this material. Students have found the exercises to be particularly interesting because they felt that the materials helped explain why they heard diatonic music the way they did. They enjoyed having an opportunity to figure out properties for themselves, rather than having everything explained to them. Some students have felt that the material helped them learn intervals and chords more solidly, and some of the more visually oriented students have noted that many of these exercises helped them see the structures more clearly when working with circle diagrams than when they first encountered these ideas only on a staff or by ear.

In more advanced work with some of these students, I have observed a significantly more receptive attitude towards the study of atonal theory. Students who were introduced to the material in this book at the beginning stages of their study of music theory found themselves able to draw upon their background to make a smooth transition to the introduction of pitch-class set theory in twentieth-century music classes, while other students who lacked this earlier training struggled more with the new concepts. Embarking on a study of diatonic set theory at the beginning stages of instruction in music theory will both enrich students' understanding of the diatonically oriented material at hand and prepare students for more advanced work.

Background

Although a number of prominent scholars have made substantial contributions to the study of diatonicism (as discussed in the For Further Study portion toward the end of this textbook), the theoretical basis of this textbook is focused primarily on two seminal articles by John Clough and his collaborators. The material in Chapter 1 stems mainly from John Clough and Jack Douthett's "Maximally Even Sets" (for full references of the scholarship discussed in this section, see the Sources Cited at the end of this textbook). Chapter 2 is based principally on material drawn from John Clough and Gerald Myerson's "Variety and Multiplicity in Diatonic Systems." Aspects of both of these articles are combined in Chapter 3. Some of the approaches taken in this textbook, as well as the definitions of relevant terms, have been adapted from these essays. Although related contributions of other scholars are introduced and fully cited in this text, ideas drawn from the two main sources cited above will appear without further acknowledgment.

My contribution to the field of diatonic set theory in this textbook is purely pedagogical. This text attempts to introduce to beginning-level students some of the innovative concepts contained in recent scholarship in diatonic theory by means of several series of interactive exercises. The theoretical concepts chosen for inclusion in this textbook coordinate especially well with a traditional approach to the study of music fundamentals. Although this book may serve as a useful but limited introduction to the field of diatonic set theory, it is intended primarily as a way to approach certain aspects of diatonic theory that are pertinent to the study of fundamentals.

Instructor Resources

The *Instructor Resources,* available online only, contains material on course planning and other information for instructors. It includes chapter abstracts, teaching notes, and suggested extensions to the material presented in the textbook. In addition, it provides supplemental exercises and solutions that are directly modeled on those in the text, blank exercise sheets that may be tailored by instructors to individual situations, and additional problems and solutions. To obtain access to these materials, please call 888-877-7240 or visit Key College Publishing on the web at www.keycollege.com.

$\mathcal{A}$CKNOWLEDGMENTS

In the mid-1990s, I began to work collaboratively with Alan Durfee, a former colleague in mathematics, to develop curricular material that describes mathematically oriented properties of the diatonic system for use in introductory music theory courses. This work was supported by a grant sponsored jointly by the National Endowment for the Humanities and the National Science Foundation. Some of the material developed for this book originated in my early work with Alan. I thank him for his valuable input in the initial formation of these pedagogical ideas.

A few years later, Dartmouth College received a National Science Foundation grant for the Mathematics Across the Curriculum (MATC) project. The MATC goal was to introduce students to applications of mathematics in a variety of disciplinary settings and to provide students with opportunities to grapple with mathematical ideas and reasoning in contexts drawn from the humanities. I was invited to include my work among the materials being developed at Dartmouth College. In addition to several core textbooks, MATC planned for a "shelf of paperbacks," a series of mini-textbooks in a variety of disciplines, all of which focus on mathematically oriented issues in non-mathematics courses. This text is one of those paperbacks. I particularly would like to acknowledge the assistance of Claude J. Poux, administrative director of the project, for coordinating the details of my contribution to MATC.

My present institution, Ithaca College, has supported the writing of this book in several ways. First and foremost, I would like to thank my Theory I students who offered many suggestions and much encouragement when encountering these materials for the first time. I am also pleased to acknowledge the support of the Center for Faculty Research and Development at Ithaca College, which provided released time for preparation of and revisions to the manuscript. Finally, I thank my colleagues for their enthusiastic receptiveness to my work, and particularly Rebecca Jemian who read and commented on earlier versions of the manuscript and the *Instructor Resources*.

I would like to express my appreciation to the staff of Key College Publishing, and in particular I thank my development editors. Cortney Bruggink, who served in this capacity during the early pre-production stages, kept the project on track, rewarded hard work with patience and flexibility, and maintained a professional yet amicable attitude throughout the process. Allyndreth Cassidy, who became the development editor as this book was heading toward production, has been enormously helpful. She has provided timely and insightful advice about both content and form, and I appreciate her earnest attention and devotion to this project. I also would like to thank the staff of Interactive Composition Corporation—especially, Brittney Corrigan-McElroy, who served as Senior Project Manager, and Erika Kauppi, who edited the manuscript.

I am indebted to Norman Carey, Eastman School of Music; David Clampitt, Yale University; and John Clough, State University of New York at Buffalo, who reviewed preliminary versions of the manuscript and provided invaluable suggestions and advice. Their perceptive observations especially helped me to expand and deepen the theoretical concepts underlying the pedagogical approaches in the text. Their enthusiasm for the project helped keep me motivated and focused when substantial portions of the text needed revision.

I give special thanks to John Clough, my mentor in graduate school at the State University of New York at Buffalo, and to whom this book is dedicated, for patiently introducing some of the profound ideas of diatonic set theory to me in the first place.

Timothy A. Johnson
Ithaca College

THE VISION OF MATHEMATICS ACROSS THE CURRICULUM

Dear Reader,

In 1994, Dartmouth College received a generous grant from the National Science Foundation to integrate mathematics throughout the undergraduate college curriculum in a five-year project, Mathematics Across the Curriculum (MATC). The project has involved over 40 faculty members from Dartmouth and various other colleges and universities representing departments of biology, chemistry, music, drama, English, art history, computer science, physics, earth science, economics, engineering, medicine, mathematics, and Spanish, producing lesson plans, short books, videotapes, and a Web site with images and text. The series of volumes published by Key College Publishing represents some of the best of the MATC collection.

These materials will make it easier for students to become more quantitatively literate as they tackle complex, real-world problems that must be approached through the door of mathematics. We hope that you, the reader, will appreciate our efforts to place the mathematics in this book completely in the context of your field of interest. Our goal is to help you see that applied mathematics is a powerful form of inquiry, and ever so much richer than mere "word problems." We trust that you will like this approach and want to explore some of the other volumes in the series.

Sincerely,

Dorothy Wallace
Professor of Mathematics
Principal Investigator: Mathematics Across the Curriculum project
Dartmouth College

INTRODUCTION

$\mathcal{D}$O YOU HAVE ANY QUESTIONS?

"Do you have any questions?" a famous composer and conductor asked an audience of music students and professors at a public lecture not so many years ago.

"Yes," replied a well-known and gifted pianist. "Why are the black and white keys of the piano arranged in that way?"

The audience sat in thought for a couple of seconds before a quiet, nervous laughter began to break the ponderous silence. Both the composer and the pianist seemed unable to arrive at a satisfying answer, but their faces showed that they were intrigued and engaged by the question.

Recent scholarship that has taken a mathematically oriented approach to diatonic musical structures has produced some of the most potentially important material on diatonicism to date. The significance of this research lies largely in its attempts to answer many of the intriguing questions that have captivated students of diatonic music for centuries. Why does the major scale seem to work so well? Why has diatonicism formed the backbone of Western music for so long—permeating both classical music of the past (and now the present) and much popular music? And, perhaps most naïve and yet apt, why *are* the black and white keys of the piano arranged as they are? Such questions continue to surface among students in introductory music theory classes, and conclusions reached in recent research in diatonic set theory may help you answer some of your own questions.

Approach

This book presents a pedagogical strategy for introducing aspects of diatonic set theory into the music theory curriculum at the beginning stages of instruction. As you begin to learn musical aspects and applications of music theory in an introductory course (or on your own), by using this book you can work simultaneously with corresponding mathematically based properties that describe aspects of and relationships within the diatonic collection. By exploring the theoretical principles behind some special aspects of the diatonic collection at an early stage, you can better understand tonal relationships between the notes of the scale and the structural significance of these relationships when encountering these ideas in your later studies.

$\mathcal{M}$ATHEMATICS AND MUSIC

The mathematics in this book is simple and direct; no previous mathematical experience is necessary. Most of the mathematical aspects of the book are conceptual rather than computational, though a few simple but useful formulas are introduced. The main orientation of the book is musical, rather than mathematical. However, approaching music fundamentals through the concepts introduced in this text will help provide you with a solid abstract foundation for musical thought based on mathematical ideas and reasoning.

Pursuing the close link between mathematics and music (as in this textbook) can transform our understanding of both, as suggested by Edward Rothstein's *Emblems of Mind: The Inner Life of Music and Mathematics*.[1] Rothstein likens the creative act of musical composition to the inspired act of constructing mathematical proofs. He finds beauty in both the musical score and the mathematical formula—in each he senses "a genius in the very notation that has developed for giving representation to ideas that seem to lie beyond ordinary language" (p. 17). Rothstein exalts mathematical ideas and musical compositions as "emblems of mind"—in which "the mind's creations can possess such mastery . . . that they can catch even the creators by surprise" (p. 4). Although the mathematics in this textbook will remain largely in the background, principles of mathematics rest at the core of every experience encountered. And examining some mathematical foundations of musical structures, in the words of Rothstein, "may lead us into profound regions we would never have stumbled on if our path were guided solely by one or the other; and our understanding of mathematics and music is bound to change based upon those connections" (p. 9).

Historical Overview

Although direct connections between number theory or group theory and music only relatively recently have begun to be explored explicitly, mathematics has been closely associated with music for centuries, primarily in the areas of tuning, temperament, and acoustics.[2] Noted musicologist Richard Crocker claimed that "whenever we undertake to explain music with integers, we necessarily begin with the simple truths the Pythagoreans set forth."[3] Although this textbook uses integers in an entirely different way than the Pythagoreans did, the simplicity of the mathematics in this book is in keeping with the flavor of the Pythagorean approach to music, which was based on simple arithmetic ratios and operations.

In the sixth century B.C.E. Pythagoras discovered that musical intervals may be obtained by means of numerical ratios between the lengths of vibrating strings. For example, strings in the ratio 2:1 produce a pure octave, 3:2 a pure fifth, and 4:3 a pure fourth. The whole tone, formed by the difference between the fifth and the fourth, is produced by the ratio 9:8 (calculated by dividing the ratios for the fifth and fourth). Linking six of these whole tones in succession, which ordinarily might be expected to be equivalent to an octave, exceeds the pure octave slightly (by a ratio called the *Pythagorean comma*).[4]

Over time, intervals have been adjusted following various mathematical schemes to compensate for the slight intervallic impurities produced by strict adherence to Pythagorean ratios.[5] For example, equal temperament (the system commonly in use in Western music today, with twelve half steps in each octave) seeks to adjust all intervals as needed to produce equal distances between all similarly placed pairs of notes. This book assumes the octave to be divided into twelve increments or half steps, but the actual tuning method adopted for these twelve increments is not a necessary part of the theories to be introduced in this text.[6] Although it is expected that you likely will be working with equal temperament, the ideas in this book would be just as effective under any other temperament involving twelve (unequal) divisions per octave. Furthermore, the theories presented in this text are also applicable to microtonal divisions of the octave (more than twelve), but such extensions to the theories will not be explored here.[7] The basic mathematical principles of Pythagoras were considerably refined and expanded by later writers, as outlined in the following brief historical survey of the association of mathematics and music.[8]

In the fourth century B.C.E. Aristoxenus conceived of and described music in terms of spatially oriented principles of geometry, rather than using an arithmetic approach based on string lengths, and he derived intervals based on spatial distances between notes rather than numerical proportions.[9] Ptolemy, in the second century (C.E.), sought to counteract the imperfection of the human perception of sounds by using precise measurement and mathematical reasoning, again based on string lengths. In the sixth century, Boethius attempted to produce a comprehensive account of ancient sources of music theory, and he transmitted arithmetic, geometric, and physical concepts of sound.

Later writers—such as Walter Odington (fourteenth century), Franchinus Gaffurius (fifteenth and early sixteenth century), and Gioseffo Zarlino (sixteenth century), among others—tried to balance mathematical approaches to musical sounds with their own perceptions of music practice. The primary issue was how take certain intervals that the ancients held as dissonant, based on mathematical reasoning, and reinterpret them as consonances based on their use in contemporaneous musical compositions—without entirely giving up the mathematical models. By the early seventeenth century, prominent scholars began to question the Pythagorean ratios.[10] Galileo Galilei, among others, determined that the ratios between the number of *vibrations* different sound sources produced, rather than the string lengths themselves, were directly responsible for the formation of musical intervals, setting the stage for further advancements in the scientific quantification of sound.

In the eighteenth century, the influential theorist Jean-Jacques Rameau derived the major triad, the primary sonority of the music of his time (and in many circles, the primary sonority to this day), from ratios of string vibrations. Although others of his day followed similar paths, "Rameau was perhaps the best known Enlightenment figure who strove to account for music in terms of mathematics and the observation of natural phenomena."[11] In the nineteenth century, working from a similar beginning point but taking mathematical and scientific aspects of music to an entirely new level of rigor, Hermann Helmholtz pioneered the study of physical and physiological acoustics.[12] Helmholtz provided a detailed study of the human ear and its role in the perception and interpretation of musical sounds and their combinations, and thereby connected physical and physiological acoustics with musical science and aesthetics. Acoustics continues to be an important part of the study of the relationship between mathematics and music, and students can benefit greatly from a course, or at least self study, in acoustics.[13]

The interrelationship between mathematics and music since the middle of the twentieth century has focused largely on twelve-tone and atonal music. Milton Babbitt, a highly influential composer and theorist, was one of the first to systemize aspects of this musical repertory by means of mathematical constructs. Many other scholars have continued in this vein, principally Allen Forte, David Lewin, and Robert Morris, to name just a few. Working from this tradition, but applying similar methods to more traditional musical resources and structures, John Clough and other scholars recently began to explore diatonic music through mathematically oriented concepts and procedures, as outlined in the For Further Study section at the end of this book. It is this more recent development in the history of the interrelationship between mathematics and music that this book explores—rather than the older and perhaps more familiar ideas and methods associated with tuning, temperament, and acoustics.[14]

How TO USE THIS BOOK

Unlike many textbooks, in this book all of the exercises provided are subsequently solved for you. You are expected to complete each exercise then compare your results with the solutions provided; the text includes a full discussion of these solutions. Alternate solutions are displayed, detailed explanations of how the exercises can be solved are offered, and the significance of the solutions are pondered. The exercises developed for this book are designed to lead you to discover principles of diatonic set theory for yourself. Rather than introducing the ideas of diatonic theory formally, then following this exposition of ideas with examples and exercises for further exploration, the exercises presented here are designed to help you to reach your own conclusions about the structure of diatonicism based on your own observations and without prior knowledge of the underlying theoretical concepts.

As you become familiar with scales, intervals, keys, and chords, the introduction of general principles underlying the structure of diatonicism in this book may give you a broader context in which to contemplate these musical building blocks. In all of these exercises, by building various patterns and structures according to prescribed stipulations, you will be able to observe many underlying abstract constructions independently. Simultaneously, you will obtain practice using the musical materials that are customarily introduced in beginning music theory classes. All of this material presents "cutting-edge" research in music theory in a non-threatening and useful way at the introductory level, and also introduces applications of mathematics that appear naturally in an introductory music theory setting.

SPATIAL RELATIONS AND MUSICAL STRUCTURES

1

SPATIAL RELATION PUZZLES

We begin our study of musical structures by considering several puzzles pertaining to spatial relations. For now, ignore any potential musical applications; concentrate only on the spatial relation problems presented. We will consider the correspondences with musical structures later in the chapter.

Placing Two, Three, Four, and Five Dots on Circle Diagrams

Exercise 1.1 contains four circles, each crossed by twelve equally spaced, short lines. In the sample, two dots have been placed on the crossing lines such that the dots are spread out as much as possible around the circle. In the same way, place three dots on the crossing lines around the second circle so that the dots are spread out as much as possible. Continue this procedure by placing four and five dots, respectively, on the crossing lines around the other two circles.

EXERCISE 1.1 Place three, four, and five dots on the crossing lines around the circles so that the dots are spread out as much as possible. The first one is done for you.

Sample: 2 dots   **a.** 3 dots   **b.** 4 dots   **c.** 5 dots

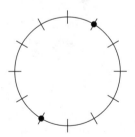 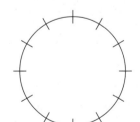 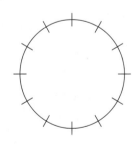

Placing three or four dots around the circles probably was easier than placing five dots; hence, we will explore the solutions to these problems first. Because there are twelve equally spaced lines crossing the circles, three or four dots can be placed around a circle without compromising the position of the dots relative to each other. To ensure that the dots are "spread out as much as possible," simply count the lines between each placed dot. However, there are several correct solutions for each of these problems, though all of these solutions are rotations of a single pattern for each problem. The number of correct solutions, or distinct rotations of the pattern of dots, can be calculated by means of a simple formula.

Greatest Common Divisor

For this formula, we will call the number of lines *crossing* the circle *c*, and the number of *dots* placed around the circle *d*. The number of distinct solutions to each problem is equivalent to the number of lines crossing a circle (c) *divided by* the greatest common divisor (GCD) of c (the number of lines) and d (the number of dots).

$$\frac{c}{\text{GCD of } (c, d)}$$

The greatest common divisor of a pair of numbers is the largest number that can divide *both* numbers evenly (with no remainder).

For the problem with three dots, the greatest common divisor of 12 (lines) and 3 (dots) is 3, because 3 is the largest number that will divide evenly into both 12 and 3. Plugging these numbers into the formula reveals the number of distinct solutions to the problem—the number of crossing lines around the circle (12) divided by the greatest common divisor of the number of lines (12) and dots (3).

$$\frac{c}{\text{GCD of } (c, d)} = \frac{12}{\text{GCD } (12, 3)} = \frac{12}{3} = 4$$

Thus, there are four distinct solutions to this problem. You can see these solutions easily by rotating an evenly spaced, three-dot pattern four times (that is, rotating the dots one crossing line to the right, or clockwise, each time), as shown in Solution 1.1a. Rotating the dots a fifth time would produce the same arrangement of dots with which you started; therefore, there are only four distinct solutions, as calculated in the formula. Your solution to placing three dots around the circle in Exercise 1.1a should match one of these provided solutions. If it does not, make corrections to your three-dot diagram, and revise your other diagrams as necessary, based on this approach. It is important that you make corrections to your own circles as you work through this book, because we shall return to these diagrams later.

SOLUTION 1.1a The four distinct solutions to placing three dots around a circle with twelve crossing lines

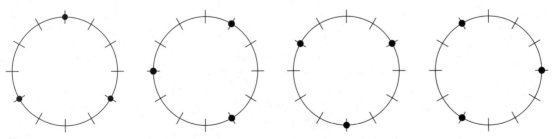

The number of correct solutions to placing four dots around a circle so that the dots are spread out as much as possible may be calculated in the same way. The greatest common divisor of 12 (lines) and 4 (dots) is 4, because 4 is the largest number that will divide evenly into both 12 and 4. Consequently, the number of crossing lines around the circle (12) divided by the greatest common divisor of the number of lines (12) and dots (4) yields the number of distinct solutions (3).

$$\frac{c}{\text{GCD of } (c, d)} = \frac{12}{\text{GCD} (12, 4)} = \frac{12}{4} = 3$$

Again, you can see these three solutions easily by rotating an evenly spaced, four-dot pattern three times (that is, rotating the dots one crossing line to the right, or clockwise, each time), as shown in Solution 1.1b. Rotating the dots a fourth time would produce the same arrangement of dots with which you started; therefore, there are only three distinct solutions, as calculated in the formula. If your solution to placing four dots around the circle in Exercise 1.1b does not match one of these solutions, make any necessary adjustments to your answer.

The three distinct solutions to placing four dots around a circle with twelve crossing lines

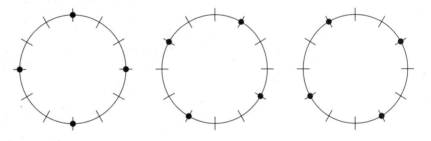

Compromises Needed to Place Five Dots

As mentioned previously (and as you, no doubt, are well aware from your own efforts), placing five dots around a circle so that the dots are spread out as much as possible is more challenging. Compromises must be made in terms of the position of the dots relative to each other. You cannot simply count an equal number of lines between each placed dot, as was possible in the circles with two, three, and four dots. Figure 1.1 shows three hypothetical attempts to place five dots around a circle. In Figure 1.1a, the *clusters* of dots are spread out from each other, but the dots in each cluster are not spread out. In Figure 1.1b, each successive dot is placed twelve-fifths (or two and two-fifths) of the way around the circle. In this way the dots are evenly dispersed around the circle, but unfortunately without regard to the crossing lines, as directed. However, moving these dots to the nearest crossing lines (or "rounding off" these dots) produces the diagram shown in Figure 1.1c, the desired response.[1] This solution (Figure 1.1c) exhibits the best compromise in terms of placing the dots so that they are spread out as much as possible. Each of the dots has at least one extra space next to it; dots with two empty spaces between them are placed as far from each other as possible.

Figure 1.1 Some hypothetical ways to place five dots around a circle with twelve crossing lines (a and b show incorrect attempts; c is a correct response)

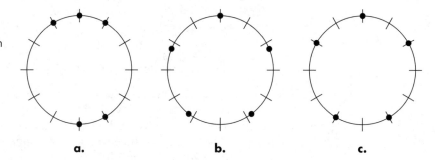

a.   b.   c.

Rotations of a Five-Dot Circle

Although Figure 1.1c shows a correct solution to placing five dots around a circle, there are several other correct solutions for this problem, all of which are rotations of this single pattern. The number of correct solutions to placing five dots around a circle so that the dots are spread out as much as possible can be calculated in the same way as with three and four dots. The greatest common divisor of 12 (lines) and 5 (dots) is 1, because 1 is the largest (and only) number that will divide evenly into both 12 and 5. Consequently, the number of crossing lines around the circle (12) divided by the greatest common divisor of the number of lines (12) and dots (5) yields the number of distinct solutions (12).

$$\frac{c}{\text{GCD of } (c, d)} = \frac{12}{\text{GCD } (12, 5)} = \frac{12}{1} = 12$$

Again, you can see these twelve solutions easily by rotating an evenly spaced, five-dot pattern twelve times (that is, rotating the dots one crossing line to the right, or clockwise, each time), as shown in Solution 1.1c. Each of these rotations produces a distinct pattern; none overlaps with any other five-dot pattern. Therefore, there are twelve distinct solutions, as calculated by the formula. If your solution to placing five dots around the circle in Exercise 1.1c does not match one of the rotations shown in Solution 1.1c, try the exercise again, in light of our discussion of making the best compromise.

The twelve distinct solutions to placing five dots around a circle with twelve crossing lines

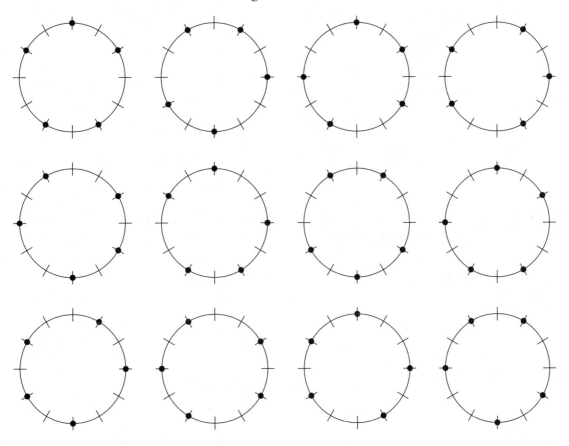

Placing Six, Seven, and Eight Dots on Circle Diagrams

Before we consider the musical importance of the diagrams you have constructed, we will continue with three more spatial relation puzzles. By now you likely will be able to solve these problems more easily, and you can predict how many possible correct solutions there are for each problem. Exercise 1.2 contains three more circles, each with twelve crossing lines. Place six, seven, and eight dots around the respective circles so that the dots are spread out as much as possible. Next, in Exercise 1.3 use the formula as before to determine how many correct solutions are possible for each circle.

E X E R C I S E 1.2   Place six, seven, and eight dots on the lines crossing the circles so that the dots are spread out as much as possible.

a. 6 dots   **b.** 7 dots   **c.** 8 dots

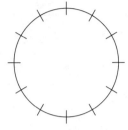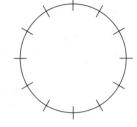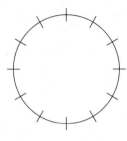

E X E R C I S E 1.3   Solve each formula to determine how many correct solutions are possible for each circle with the indicated number of dots.

a. 6 dots $\dfrac{c}{GCD(c, d)} = \dfrac{\quad}{GCD(\ ,\ )} = \dfrac{\quad}{\quad} = \boxed{}$ correct solutions

b. 7 dots $\dfrac{c}{GCD(c, d)} = \dfrac{\quad}{GCD(\ ,\ )} = \dfrac{\quad}{\quad} = \boxed{}$ correct solutions

c. 8 dots $\dfrac{c}{GCD(c, d)} = \dfrac{\quad}{GCD(\ ,\ )} = \dfrac{\quad}{\quad} = \boxed{}$ correct solutions

As before, placing dots around two of these circles probably was easier than placing dots around the other one, because only one of these problems requires compromises in spreading out the dots. Placing six dots around the circle can be accomplished simply by placing a dot on every other crossing line, skipping one line each time. Also, as demonstrated in Solution 1.2a and as calculated in Solution 1.3a, there are only two correct solutions to this problem; the pattern of dots can be rotated only one place to the right without duplicating the original pattern.

Because placing seven dots around the circle is more challenging, we will consider the eight-dot spatial relation puzzle next. The solution to placing eight dots around the circle is relatively unproblematic. However, instead of an even arrangement of dots and spaces—as in the circle diagrams with two, three, four, and six dots—pairs of adjacent dots alternate with single spaces, as shown in Solution 1.2c. Three distinct solutions are possible with eight dots, as calculated in the formula shown in Solution 1.3c, and as illustrated in Solution 1.2c. Your answer to Exercise 1.2c should match one of these three circle diagrams.

The Complement (Eight and Four Dots)

Note that the *crossing lines that are free of dots* in each eight-dot circle diagram (Solution 1.2c) have the same arrangement as the *dots* in the four-dot problem explored earlier (Solution 1.1b). Likewise, the dots in each eight-dot diagram have the same arrangement as the crossing lines that are free of dots in the four-dot problem. Thus, in each case the dots are spread out as much as possible, and the lines without dots are spread out as much as possible as well. This special relationship between the four-dot circles and the eight-dot circles is called a *complement*. The complement completes the circle; in this case, the arrangement of dots in a four-dot diagram would complete the circle in an eight-dot diagram by filling in the empty lines around the circle. In other words, if you superimpose a four-dot circle over an eight-dot circle and rotate the two diagrams properly, the dots and empty lines will match up. The dots on the four-dot circle will appear directly over lines without dots on the eight-dot circle, and the dots on the eight-dot circle will appear directly under the lines without dots on the four-dot circle.

Compromises Needed to Place Seven Dots

Placing seven dots around a circle so that the dots are spread out as much as possible poses a similar challenge to the problem of placing five dots around a circle. Because there is no way to disperse the dots around the crossing lines evenly, some compromises are necessary to complete the problem. Figure 1.2 illustrates the nature of the problem by giving a few hypothetical solutions. Because placing six dots around the circle was easily accomplished, as shown in Solution 1.2a, this experience might suggest the diagram in Figure 1.2a—where the six dots are spread out as much as possible, and a single dot remains to be placed. On one hand, it may appear that the best compromise, therefore, is to place the leftover dot on one of the remaining lines, as in Figure 1.2b. However, this arrangement clumps three dots together. On the other hand, perhaps the diagram in Figure 1.2c would improve the situation because the two pairs of adjacent dots are close together, somewhat resembling the solution to the five-dot problem shown in Solution 1.1c. However, the circle still seems unbalanced. Clearly, defining "spread out as much as possible" is the issue here, but we will delay any formal definition of this concept until later in the chapter and rely more on intuition for the moment. An analogy might be helpful in solving this thorny issue.

Figure 1.2 Some hypothetical ways to place seven dots around a circle with twelve crossing lines (a shows the problem; b and c show incorrect attempts; d is a correct response)

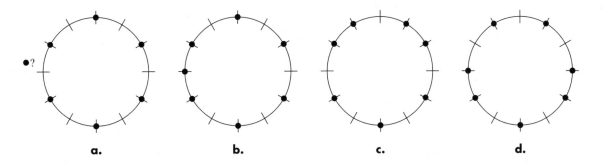

a.   b.   c.   d.

The two distinct solutions to placing six dots around a circle with twelve crossing lines

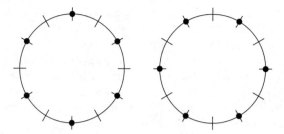

The twelve distinct solutions to placing seven dots around a circle with twelve crossing lines

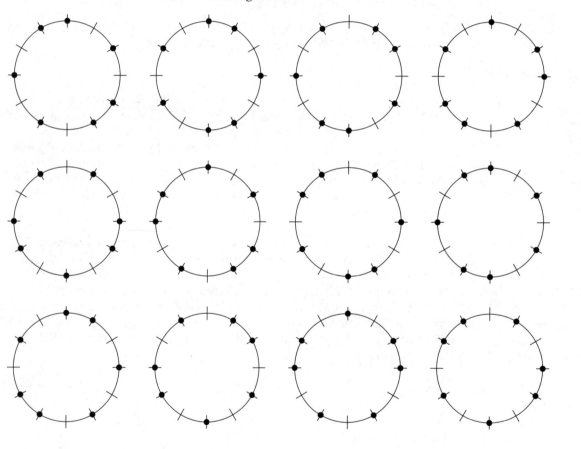

The three distinct solutions to placing eight dots around a circle with twelve crossing lines

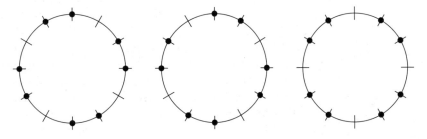

Calculating the number of possible solutions for placing dots around circles

a. 6 dots $\dfrac{c}{\text{GCD}(c,\, d)} = \dfrac{12}{\text{GCD}(12,\, 6)} = \dfrac{12}{6} = \boxed{2}$ correct solutions

b. 7 dots $\dfrac{c}{\text{GCD}(c,\, d)} = \dfrac{12}{\text{GCD}(12,\, 7)} = \dfrac{12}{7} = \boxed{12}$ correct solutions

c. 8 dots $\dfrac{c}{\text{GCD}(c,\, d)} = \dfrac{12}{\text{GCD}(12,\, 8)} = \dfrac{12}{8} = \boxed{3}$ correct solutions

The Dinner Table Analogy

Suppose you have a round dinner table surrounded by twelve evenly dispersed chairs, none of which can be removed.[2] You have invited seven guests to dinner and want guests to be spread out as much as possible around the table, so that no one is left out of the conversation (by being too far isolated from other guests) and so that no individual group or groups monopolize the conversation (by being too close together compared to the other guests). The circle diagrams in Figure 1.2 can be used to illustrate this dining dilemma. With the arrangement shown in Figure 1.2a, six of the guests are happily chatting away, while one unfortunate guest is left to walk around the table (fine for a waiter, but not a guest!). With the arrangement depicted in Figure 1.2b, in which the roving guest has sat down, a single group of three people have formed a clique, while the others seem to have been excluded from their juicy gossip. Although the arrangement shown in Figure 1.2c rectifies the clique problem, the bulk of the conversation still seems to be occurring on a single side of the table. Because arranging seven guests evenly around a twelve-seat table requires at least two pairs of people sitting in adjacent chairs, perhaps a more convivial conversation would be encouraged if the two pairs were placed as far apart as possible, as in Figure 1.2d. In this way, everyone can share equally in the discussion, and the adjacent pairs of guests are placed as far from each other as the seating arrangement allows. This analogous solution achieves the original goal of placing the dots around the circle so that the dots are spread out as much as possible. One might suggest seating all seven guests in a row of adjacent chairs; however, this arrangement gives a distinct disadvantage to the persons on each end of the row, who must come away from the dinner with severe neck strain from always leaning in a single direction to participate in the merrymaking.

This analogy works in all of the puzzles, no matter how many dots are placed around the circles. However, the analogy is most useful, perhaps, with the more problematic puzzles—how to place five or seven dots. In addition, the strategy discussed in connection with Figure 1.1b, where five dots were evenly dispersed around the circle then adjusted (rounded) to the nearest lines, also works for seven dots, as shown in Figure 1.3. Figure 1.3a shows the dots evenly dispersed around the circle, but without regard to the crossing lines. Each dot is placed twelve-sevenths (or, one and five-sevenths) of the way around the circle, beginning (arbitrarily) at the bottom of the circle. Moving these dots to the nearest crossing lines (or "rounding off" these dots) produces the diagram in Figure 1.3b, the desired response. This solution exhibits the best compromise in terms of placing the dots so that they are spread out as much as possible, and it is the same as the solution given in Figure 1.2d and discussed in conjunction with the dinner table analogy.

Figure 1.3 Two possible ways to place seven dots around a circle with twelve crossing lines (a shows an incorrect attempt; b is a correct response)

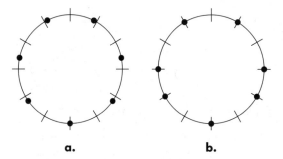

a.   b.

As with the five-dot circle, there are also twelve correct solutions to the problem of placing seven dots around a circle, as calculated in Solution 1.3b. All of these possible answers, which are each a rotation of a single pattern of dots, are given in Solution 1.2b—including the solutions shown in Figures 1.2d and 1.3b.

The Complement (Seven and Five Dots)

In addition to their more obvious problem-solving similarities, there is also a complementary relationship between the seven-dot and five-dot circles. The crossing lines that are free of dots in each seven-dot circle diagram (Solution 1.2b) have the same arrangement as the dots in the five-dot solution shown earlier (Solution 1.1c). Likewise, the dots in each seven-dot diagram have the same arrangement as the crossing lines that are free of dots in the five-dot problem. Put another way, the five-dot circle diagram superimposed over the seven-dot circle diagram will complete the circle by filling in all of the lines around the circle with dots.

Previously, you were asked to arrange dots around crossing lines and ignore how these diagrams might be used musically. These spatial relation puzzles were presented in this way to allow you to actively explore these abstract constructions firsthand and in detail. However, the *musical* application of these exercises is our main concern, and we will turn to the consideration of this aspect now.

$\mathcal{M}$USICAL STRUCTURES FROM GEOMETRIC FIGURES

We now will attempt to determine what musical structures are related to the distinct arrangements of dots that are spread out as much as possible around the circles. Each pattern of dots corresponds to a familiar musical structure, and most of these structures may be determined by students who are willing to experiment independently with various possible orderings and who are able to recognize these musical patterns by ear. Some of these musical structures may be more advanced, depending on your current level of study, but you are encouraged to try to identify all of the patterns as well as you can, even if you lack the proper terminology. We will discuss all of the solutions after you have made your best attempts.

Labeling Notes on Circle Diagrams

To complete Exercise 1.4, you will label lines around circles with musical note names, play various patterns on a piano and listen to the resulting sounds, and attempt to identify the musical structures produced. Using your corrected diagrams from Exercises 1.1 and 1.2, assign note names to the twelve crossing lines, ascending chromatically around each circle in a clockwise manner. Use either sharps or flats (but not both) for all chromatic notes. For example, in the two-dot diagram in Exercise 1.1, label any one of the crossing lines as "C." Then, name the next line (moving clockwise) as "C♯," the next line "D," the next line "D♯," and so forth, ending with the last crossing line labeled "B" (next to the "C" with which you began). (I am using sharps arbitrarily here—you could complete the same labeling procedure using flats instead of sharps.) For this exercise, it does not matter where you begin to label the lines. Because there are twelve crossing lines around the circles and twelve chromatic notes in an octave, no matter where you start, you will label all of the lines with the notes C through B to complete the octave of note names. In the same way, label all of the lines around each circle of your corrected diagrams in Exercises 1.1 and 1.2.

Circle Diagrams and the Piano

Next, take your circle diagrams to a piano, and play the notes corresponding to the dots for each circle. Play the notes assigned to the dots consecutively as you move clockwise around the circle. Try starting on different notes—that is, at different positions on the circle—when you play (still corresponding to the same dots and note names on the circle but *beginning* on different dots, or notes on the piano). For some of these patterns, you may not be able to recognize the musical structures unless you *begin* playing the pattern on one particular note—so try them all.

Listen and try to recognize what musical structures you are playing. You may wish to consider the note names you used to label the dots, but the actual spelling of these musical structures on the circle diagrams may not always correspond directly with musical practice, in terms of enharmonic equivalency, because we are using only sharps in these diagrams. (Different spellings of a note—such as C♯ and D♭, D♯ and E♭, and B♯ and C—are called *enharmonically equivalent*. We are using sharps in the sample, but the enharmonically equivalent flats may be substituted as needed to facilitate recognition of the musical structures depicted.) It might help to plot each pattern on a staff, but rely primarily on your ear as you play the patterns on the piano. Your goal is to identify the musical structure corresponding to each pattern of dots that are spread out as much as possible around the circles. For example, the two-dot pattern corresponds to a musical interval, the three-dot pattern forms a triad, the seven-dot pattern produces a familiar scale (starting on the "right" note might help you recognize it), and so forth. Do as many as you can before reading ahead, then we will discuss the solutions in detail. Also, be diligent in your efforts, because our focus in this exercise is on your own independent discovery of these musical structures. You can record your answers in Exercise 1.4.

EXERCISE 1.4   Using the circle diagrams you constructed in Exercises 1.1 and 1.2, assign note names to the twelve lines chromatically around the circles in a clockwise manner, using either sharps or flats (but not both) for all chromatic notes. What musical structures are formed for each circle by the notes corresponding to the dots? Play the notes on a piano to help you determine the musical structures; try starting on different notes.

The circle with: *produces a(n):*

2 dots _____ (interval)

3 dots _____ (triad)

4 dots _____ (seventh chord)

5 dots _____ (scale)

6 dots _____ (scale)

7 dots _____ (scale)

8 dots _____ (scale)

Checking Note Labels on Circle Diagrams

Figure 1.4 shows one of many possible ways to label the circle diagrams with note names. Your own diagrams may vary both in terms of where the dots are placed around the circle—as suggested by the many possible solutions given in Solutions 1.1 (a–c) and 1.2 (a–c)—and which lines correspond to which notes. However, regardless of how you label the lines around your circles, you can compare your own work with Figure 1.4 by rotating the circles until the note names correspond. In this way, you also can verify your own labeling schemes. The identities of the musical structures formed will be constant, provided that your note names ascend chromatically through a single octave moving clockwise around the circle, even if the dots in your solutions correspond to different notes than those shown in the figure.

Figure 1.4 Note names corresponding to circles with two through eight dots

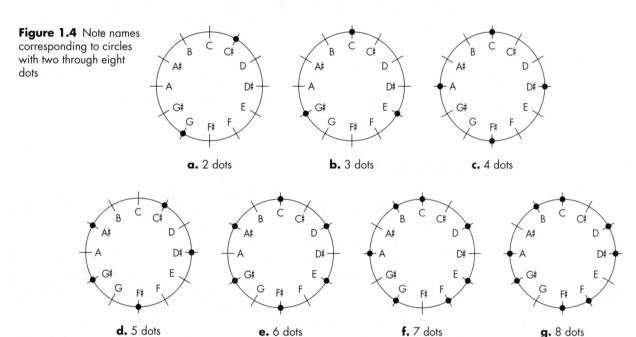

a. 2 dots   **b.** 3 dots   **c.** 4 dots

d. 5 dots   **e.** 6 dots   **f.** 7 dots   **g.** 8 dots

Solution 1.4 summarizes the correct answers to this exercise. Your answers may vary slightly from the given solutions, depending on how you labeled the notes and how specific you were in your identifications. Also, recall that some of these solutions may be more advanced than your present level of study. The following discussion will help you appreciate each of these musical structures, even if you are not yet aware of some of their names.

The Tritone

The pattern of dots on the circle with two dots corresponds to the interval of a tritone. You may have identified this interval instead as either an augmented fourth or a diminished fifth, but both of these specific intervals can be identified simply as *tritones*—meaning three whole steps, or the equivalent. The circle diagram (Exercise 1.1, Sample) clearly depicts why these two intervals, augmented fourth and diminished fifth, *sound* exactly the same due to the even placement of notes (or dots) around the circle. If you consider the interval beginning with one of the dots, then the interval produced is an augmented fourth; on the other hand, beginning with the other dot will yield a diminished fifth. Yet both intervals project the equivalent of six half steps or three whole steps, as is clearly depicted by the crossing lines on the circle diagram but is not evident in musical notation on a staff.

The musical structures formed for each circle by the notes corresponding to the dots

The circle with: *produces a(n):*

2 dots	tritone	(interval)
3 dots	augmented triad	(triad)
4 dots	diminished seventh chord	(seventh chord)
5 dots	pentatonic scale	(scale)
6 dots	whole-tone scale	(scale)
7 dots	diatonic scale (major, minor, etc.)	(scale)
8 dots	octatonic scale	(scale)

The Augmented Triad

The circle with three dots produces an augmented triad. You may have labeled a particular note as the root for this triad, but any root is possible, depending on your configuration of notes. Your answer may or may not be inverted, depending on how you labeled the crossing lines on the circle with notes, and depending on the note with which you began. However, no matter how an augmented triad is inverted, it still can be interpreted as a root position augmented triad by sound alone (disregarding enharmonic spellings). The evenly spaced distances between each note, as depicted in the circle diagram in Exercise 1.1a, suggest that any of the three dots can serve as the root (or starting dot) of this chord. It is only the spelling (or specific note names) that designates a root. When you play the notes on the piano, you can identify the structure as an augmented triad by sound, regardless of which note you use as the starting note. In this way, you can identify the chord by ear, even if the actual spelling of the notes diverges from the pattern you would have expected.

The Diminished Seventh Chord

The circle with four dots yields a diminished seventh chord. Again, any root is possible for this chord, and your chord may or may not be inverted—due to differences in labeling dots and notes. As with the augmented triad, the diminished seventh chord also sounds essentially the same in all inversions, enharmonic spellings notwithstanding. Your circle diagram in Exercise 1.1b, with its evenly spaced dots and notes, beautifully illustrates this advanced musical principle as well. Because all of the notes are equidistant from one another, it makes no difference which note is considered the root. The four notes will always sound like a diminished seventh chord, regardless of how these notes are distributed relative to each other.

The Pentatonic Scale

The circle with five dots has a less obvious solution, because the dots are not evenly dispersed around the circle, as discussed earlier in this chapter. The dots are spread out as much as possible around the circle diagram in Exercise 1.1c, but compromises were made to accommodate all five dots. This spatial pattern corresponds to a pentatonic scale. (Literally, a pentatonic scale suggests any scale of five notes, but in traditional musical discourse only this specific interval pattern is associated with the term *pentatonic* in most cases.) This scale is most familiarly formed by using the black keys on the piano. However, any notes corresponding to the dots in this circle diagram will yield a pentatonic scale. At the piano, compare the sound of the notes as you labeled them in your circle diagram with the scale pattern using only black keys. Try starting on different notes in playing these scales until the two scales sound similar.

The Whole-Tone Scale

The circle with six dots produces a whole-tone scale. As suggested by its name, the whole-tone scale consists of only whole steps. These whole steps can be seen easily in the circle diagram in Exercise 1.2a by the single-spaced gaps between each of the dots. Because you have labeled each line consecutively through the chromatic scale, the distance between each adjacent line is equivalent to a half step. Therefore, each pair of adjacent dots in the diagram is separated by the interval of a whole step (or two half steps). As with the other scales associated with evenly dispersed patterns of dots, the whole-tone scale sounds the same no matter which note is played first in the scale. Hence, it is difficult to hear where this scale begins and ends. Play the whole-tone scale associated with your labeled circle diagram slowly over several octaves, both ascending and descending. Slow down occasionally or even stop momentarily, and notice how any note could serve as an effective conclusion to the scale, regardless of which note you designated as the beginning or tonic note. Later we will observe how this scale sounds compared to the scale formed by the seven-dot circle.

The Diatonic Collection (Major Scale)

The pattern formed by the circle with seven dots in Exercise 1.2b yields a diatonic collection—or more familiarly, a major scale—if you begin with the right note. Beginning with another particular note produces a natural minor scale. Furthermore, depending on which note you play first in your scale, the pattern corresponds to each of the seven modes—ionian, dorian, phrygian, lydian, mixolydian, aeolian, and locrian. The tonic notes of these scales and modes will vary depending on how you label the lines around the circle with note names; however, all of these structures are formed from this arrangement of notes/dots, regardless of the labels—only the tonics change. Figure 1.5 illustrates how each of these scales and modes can be formed using a seven-dot circle.

Figure 1.5 The scales and modes that can be formed using a seven-dot circle

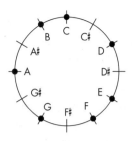

Play each of these scales on a piano, and listen for the subtle changes in the order of whole and half steps associated with each one. Observe that the notes used are the same in each scale/mode; only the tonic notes for the scales/modes are different. Thus, the collection of notes used and the relationships between adjacent notes are fixed. The *diatonic collection* is a term that generally identifies this particular arrangement of notes, regardless of tonic, or starting note.

In this book I use the term *diatonic* in its restrictive sense, to refer to the pattern of whole and half steps that corresponds to the white keys of a piano, or transpositions of this pattern. Other sources and contexts sometimes use the term more loosely to include other seven-note collections as well, such as the harmonic minor and the ascending melodic minor. Eytan Agmon and other music theorists have defined *diatonicism*, or a "diatonic tone-system," precisely based on some of the same principles discussed in this text and other similar constructs.[3] In this book, I use the term *collection* instead of *scale* when no tonic note or scalar ordering is implied. Using this term is similar to the approach we took with the augmented triad and diminished seventh chord, discussed earlier, where we ignored root and inversion, and instead named these structures more generally. Thus, the *diatonic collection* is an unordered group of notes with a fixed arrangement of whole and half steps (but including rotations of that arrangement); each of the scales shown in Figure 1.5 are diatonic in this strict sense. The other musical structures discussed, such as pentatonic and whole tone, also can be referred to as collections rather than scales. These constructs will be referred to as *collections* when generality is implied, but in other cases the more familiar term *scale* will be retained when order is implied or when no ambiguous meaning will result. The term *collection* is particularly desirable for the diatonic because all of the scales and modes connected with this collection are prevalent in musical discourse.

Play the diatonic collection associated with your labeled circle diagram slowly over several octaves, both ascending and descending, as you did with the whole-tone scale. Listen carefully for the pattern of whole and half steps associated with this collection. Note that, although slowing down and stopping on any particular note might help you to perceive that note as the tonic of a particular scale or mode, the uneven intervallic distances between adjacent notes have quite a different effect on how you perceive this collection, as opposed to the whole-tone scale where all intervallic distances between adjacent notes are equal. The fixed location of the half steps in the diatonic collection helps the listener to remain oriented within a scale from octave to octave, which is not possible with the intervallically consistent whole-tone scale.

The Octatonic Scale

Finally, the circle with eight dots in Exercise 1.2c yields a scale with an unchanging pattern that is in some ways similar to the whole-tone scale described earlier. This eight-note scale is called the *octatonic scale*. As with the five-note pentatonic scale, it is named for the number of notes it contains—eight. The traditional interval pattern associated with this scale corresponds to the structure formed by this particular eight-dot pattern, with alternating whole and half steps. The octatonic scale can begin with any note, and it can open intervallically with either a whole step or a half step.

Another name for this scale is the *diminished scale,* the preferred label in jazz circles, because if we take every other note of the scale, a diminished seventh chord is the result. For example, beginning with the C at the top of the diagram, if we take every other dot in Figure 1.4g (C, D♯, F♯, A), a diminished seventh chord results (in this case, a D♯ diminished seventh chord). From another perspective, this property of the octatonic scale also can be observed by superimposing the four-dot circle diagram over the eight-dot circle diagram. Because the four-dot circle corresponds to the diminished seventh chord, as discussed earlier, any rotation of this circle where the dots overlap the dots of the eight-dot circle will show where a diminished seventh chord can be formed from the notes of the octatonic scale.

Quantifying Musical Structures

Now that we have examined the musical structures that correspond to the various ways to place dots around circles so that the dots are spread out as much as possible, we return to the idea presented earlier concerning the number of ways that each of these patterns of dots can be formed. Recall the formula for determining the number of distinct solutions to the spatial relations puzzles posed earlier in this chapter. Exercise 1.5 offers an opportunity to verify *musically* the results that we obtained using that formula and that we observed in the various rotations of the circle diagrams. You can use either sharps or flats for chromatic notes in this exercise, depending on which marking best seems to represent the musical structure you are forming. Therefore, be careful with enharmonically equivalent notes, which are considered identical in this exercise because they will correspond to the same dot on a circle. Also, notice the effect of octaves in this exercise: The dots around the circles indicate note names only and suggest nothing about what octave to use. Thus, two notes an octave apart are considered equivalent in this case. Two of the examples (c and d) have been completed for you as samples; some parts of others also have been solved.

Enharmonic and Octave Equivalence

Exercise 1.5c, which is solved for you, provides an excellent example of the issues of enharmonic and octave equivalence. The first diminished seventh chord shown (C–E♭–G♭–B♭♭) has the same notes as the last diminished seventh chord shown (D♯–F♯–A–C). Here the Cs are an octave apart (therefore octave equivalent), and the other three pairs of notes are enharmonically equivalent to each other (E♭ = D♯, G♭ = F♯, B♭♭ = A). Keep this sample in mind as you complete the exercise, and rely on the formula to determine how many distinct forms of the musical structure you should be able to form.

First, determine the number of distinct transpositions of each structure by completing the given formula. Then, for each musical structure, transpose the structure repeatedly up by half step until the transposed structure has the same notes as the initial musical structure (regardless of octave or enharmonic spellings).

$$\frac{c}{GCD(c, d)} = \text{number of distinct ways to form the musical structure}$$

a. 2 dots, tritone

$$\frac{12}{GCD(12, 2)} = \frac{12}{2} = 6 \text{ distinct tritones}$$

b. 3 dots, augmented triad

$$\frac{}{GCD(\ ,\ )} = \frac{}{} =$$

c. 4 dots, diminished seventh chord

$$\frac{12}{GCD(12, 4)} = \frac{12}{4} = 3 \text{ distinct diminished seventh chords}$$

d. 5 dots, pentatonic scale
$$\frac{12}{GCD(12, 5)} = \frac{12}{5} = 12 \text{ distinct pentatonic scales}$$

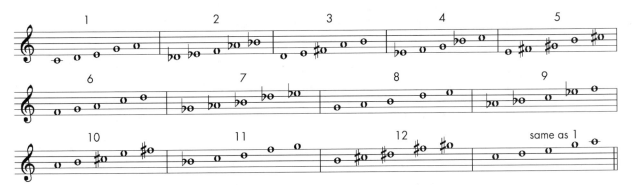

e. 6 dots, whole-tone scale
$$\frac{}{GCD(\ ,\ )} = \frac{}{} =$$

f. 7 dots, major scale
$$\frac{12}{GCD(12, 7)} = \frac{12}{7} = 12 \text{ distinct major scales}$$

g. 8 dots, octatonic scale
$$\frac{}{GCD(\ ,\ )} = \frac{}{} =$$

Checking Quantities of Musical Structures

After you complete your work, play all of the musical structures in parts a through g on a piano. Then check your answers against Solution 1.5. Your answers may vary in terms of starting points and enharmonically equivalent notes, but otherwise should match the solutions provided. As verified by completing this exercise, the result of each formula matches the number of chromatically transposed musical structures formed by the corresponding circle diagram: six tritones, four augmented triads, three diminished seventh chords, twelve pentatonic scales, only two whole-tone scales, twelve diatonic (in this case, major) scales, and three octatonic scales.

SOLUTION 1.5 The number of distinct transpositions of various structures solved by formula and shown on staves

$$\frac{c}{GCD(c, d)} = \text{number of distinct ways to form the musical structure}$$

a. 2 dots, tritone $\qquad \dfrac{12}{GCD(12, 2)} = \dfrac{12}{2} = 6 \text{ distinct tritones}$

b. 3 dots, augmented triad $\qquad \dfrac{12}{GCD(12, 3)} = \dfrac{12}{3} = 4 \text{ distinct augmented triads}$

(c and d are solved in Exercise 1.5)

e. 6 dots, whole-tone scale $\qquad \dfrac{12}{GCD(12, 6)} = \dfrac{12}{6} = 2 \text{ distinct whole-tone scales}$

f. 7 dots, major scale $\dfrac{12}{\text{GCD}(12, 7)} = \dfrac{12}{7} = 12$ distinct major scales

g. 8 dots, octatonic scale $\dfrac{12}{\text{GCD}(12, 8)} = \dfrac{12}{4} = 3$ distinct octatonic scales

In working with the original spatial relation puzzles and attempting to place dots around circles, I have purposely avoided defining formally this notion of "spread out as much as possible." Now that we have explored these spatial patterns and their musical significance in an informal way, we will turn to a more formal approach to this concept using a familiar musical measuring device, the interval. Our purpose here is to obtain a more precise way of determining if a musical structure is spread out as much as possible, and we will also have an opportunity to work with musical intervals in an abstract environment.

$\mathscr{A}$N INTERVALLIC DEFINITION

Maximally Even

Until now, our best way to determine if the dots around a circle are spread out as much as possible has been to use the dinner table analogy or to rely on intuition. However, a more precise definition of this idea would solidify our understanding of the concept and might prove useful in our study of other musical structures. With a formal definition of the concept, we could examine any musical structure to determine if it corresponds to the pattern of relationships exhibited by the musical structures we have been forming in this chapter. Accordingly, we will label the idea "spread out as much as possible" as *maximally even,* and we will define maximally even in terms of the distances between dots around a circle—or in a musical sense, in terms of the intervals between notes. The distances between dots around a circle will be measured in two ways, called *c distances* and *d distances*.[4] A formal definition of the term maximally even will require some new terms but will lead to a deeper understanding of the concept.

c Distances and d Distances

A *c distance* indicates the clockwise distance between dots measured in *crossing lines*. To determine the c distance between two dots, count the number of *crossing lines* from one dot to the other (but excluding the crossing line upon which the initial dot is placed). Musically speaking, a c distance corresponds directly to the number of half steps between two notes, because the distances between crossing lines around a circle represent half steps. A *d distance* indicates the clockwise distance between dots measured in *dots*. To determine the d distance between two dots, count the number of *dots* from one dot to the other (but again excluding the initial dot).

The largest possible clockwise c distance between two dots will always be $c - 1$, or the number of lines around the circle minus one. Thus, in our circles with 12 crossing lines, the largest c distance is 11, or $12 - 1$. A hypothetical c distance equal to c (12 in this case) would only return us to the starting point—essentially a c distance of zero, but distances of zero will not be counted in these exercises. And hypothetical c distances larger than c (in this case, larger than 12) would not have taken the shortest possible clockwise route between the two dots. The largest possible clockwise d distance will always be $d - 1$, or the number of dots around the circle minus one. Thus, the largest d distance for a circle with four dots is 3, or $4 - 1$.[5]

The circle diagram in Figure 1.6 illustrates these two methods of counting. Adjacent dots, bracketed at the top of the diagram, have a d distance of 1, regardless of the number of crossing lines between the dots, whereas the bottom left bracket shows a d distance of 2, because the A is two dots away from the F♯. In musical terms, a d distance of 1 refers to adjacent notes, whereas d distance of 2 skips a note. On the other hand, each c distance shown on the circle diagram indicates how many crossing lines, or half steps, separate the dots connected by the brackets. Thus, the upper right and lower left brackets in the diagram both show c distances of 3, or three half steps, though their d distances vary. Also, the two bracketed pairs of dots at the top of the diagram have respective c distances of 1 (between B and C) and 3 (between C and D#), based on the number of half steps between these notes.

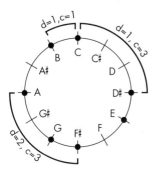

Figure 1.6 Sample procedure for counting c distances and d distances

Definition of Maximally Even

Now that we have defined c distances and d distances, we can use these terms to define the term maximally even. To determine if a circle diagram is maximally even, we will measure the clockwise distances between every pair of dots (or notes). Formally, a circle diagram is defined as maximally even if for each d distance (or distance between dots, measured in dots) there are only one or two possible c distances (or distances between dots, measured in crossing lines, or half steps). And if there are *two* c distances for a particular d distance, then the c distances are *consecutive numbers* (such as 2 and 3, 3 and 4, and so forth). In other words, to determine if a musical structure is maximally even, count the number of half steps between all pairs of notes (not just adjacent ones). For each comparably placed pair of notes (or d distance), there must be either one quantity of half steps (a single number, such as 3) or two consecutive quantities of half steps (two adjacent numbers, such as 2 and 3).

The Dinner Table Analogy is Maximally Even

In comparing the definition of maximally even with the dinner table analogy discussed earlier, we observe that the stipulations placed on the number of c distances associated with each d distance ensure that the guests (or dots) are spread out as much as possible throughout the seating arrangement (or circle). If there is only one c distance for each d distance, the distances between each similarly situated pair of guests is exactly equivalent (in terms of the number of chairs separating them). This equivalence extends not only to adjacently seated guests but also to those who are seated at some distance across the table. Thus, if only one c distance is associated with each d distance, the guests indeed are arranged in a maximally even way, because no guest is closer in proximity to any of the other guests. Furthermore, if two c distances correspond to a particular d distance, these c distances must be consecutive numbers to ensure that the separation between any two guests is as even as possible. If, on the other hand, some pairs of adjacent guests (d distance of 1) are in adjacent chairs (c distance of 1) and others are separated by two empty chairs (c distance of 3), then the spacing between similarly situated guests would be uneven—some guests would be whispering, others shouting! Likewise, if *more* than two c distances correspond to a particular d distance, then there must be a way to even out the arrangement. For example, if a row of three guests are situated one, two, and three chairs apart, why not rearrange the trio so that they are seated in every other chair? Accordingly, there must be either one or two c distances (number of chairs apart) corresponding to each d distance (number of guests apart), and if there are two c distances for a particular d distance, then the c distances must be consecutive numbers. In short, the definition of maximal evenness merely formalizes our understanding of "spread out as much as possible," as depicted in the dinner table analogy and as observed in the circle diagrams completed earlier. Moreover, through this formal definition and process, we easily can check *all* pairs of dots in a diagram to be certain that they are evenly placed with respect to each other, rather than form conclusions based solely on a visual inspection of adjacent dots.

Interval Tables

The application of the formal definition to a circle diagram can be displayed most clearly through the use of a table showing the distances between dots in terms of both the number of dots and the number of crossing lines as shown in the following examples. We begin with the four-dot circle you solved in Exercise 1.1c. Figure 1.7 contains a four-dot circle diagram (representing a diminished seventh chord) that we earlier determined is maximally even by generally observing the pattern of dots. The table beside the diagram shows that the pattern of dots is indeed maximally even according to the formal definition. As shown in the first row of the table and as illustrated in the first diagram at the bottom of the figure, each d distance of 1 (between adjacent dots) corresponds to a c distance of 3 (or three half steps). In other words, each pair of adjacent dots on the circle (d distance of 1 dot) is separated by a distance of three crossing lines (c distance of 3), as indicated by the arrows. The second row of the table shows that each d distance of 2 corresponds to a c distance of 6. The second diagram at the bottom of the figure illustrates this tabulation by showing how each pair of dots that spans a distance of two dots around the circle (d distance of 2) corresponds to a distance of six crossing lines (c distance of 6), or six half steps, as shown by the arrows. Finally, each d distance of 3 (moving three dots clockwise around the circle) corresponds to a c distance of 9 (or nine half steps), as shown on the bottom row of the table and as illustrated by the last diagram at the bottom of the figure, with c distances shown by the arrows.

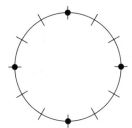

CLOCKWISE DISTANCE BETWEEN DOTS	
d distance	**c distance**
1	3
2	6
3	9

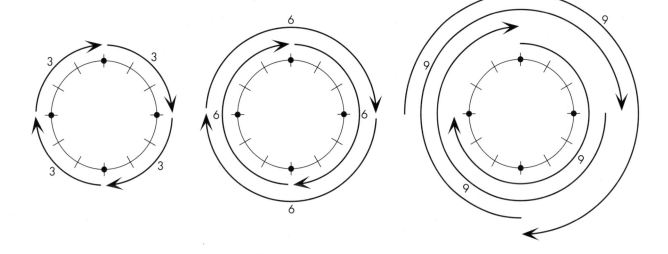

In summary, as demonstrated in the table, there is only one c distance for each d distance in the circle of four dots. Therefore, the diminished seventh chord is indeed maximally even. The last row in this table is actually redundant, because a distance of three dots moving clockwise around the circle is exactly the same as a distance of one dot moving counterclockwise. Therefore, the number of crossing lines between these dots will exactly correspond and will sum to 12, because there are a total of twelve crossing lines around the entire circle (and twelve half steps in an octave). Nevertheless, to exactly match the original definition of maximally even, we will continue to calculate these larger intervals as we check circle diagrams (and musical structures) for maximal evenness. (In musical terms, this relationship between distances is called *interval inversion*. Inversional relationships within the diatonic collection will be discussed in more detail at the end of this chapter.)

A Negative Example

Figure 1.8 contains a four-dot diagram that does not appear to be maximally even by visual inspection, and the table of distances confirms this assertion. Three different distances between dots measured in crossing lines (c distances) correspond to each distance measured in dots (d distances). For example, counting clockwise from the dot at the top of the circle, the adjacent dots in the diagram are separated by distances of 3, 3, 4, and 2 crossing lines, as illustrated in the second part of the figure. According to the definition, a circle diagram is maximally even if there are only *one or two* possible distances between dots measured in crossing lines. Therefore, the fact that three c distances correspond to a d distance of 1, even without looking at the other relationships between dots, verifies that this structure is not maximally even, as we anticipated from our initial observation of the diagram.

Figure 1.8 Checking another four-dot circle diagram to see if it is maximally even according to the definition

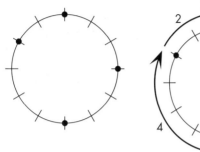

CLOCKWISE DISTANCE BETWEEN DOTS	
d distance	**c distance**
1	2, 3, 4
2	5, 6, 7
3	8, 9, 10

Testing a Five-Dot Diagram for Maximal Evenness

Figure 1.9 contains a five-dot diagram and its corresponding table. Here, for each d distance there are two c distances, and these c distances are consecutive numbers in each case. Therefore, the circle diagram shown (and the pentatonic scale, which this diagram represents) is maximally even, as we would have expected from our discussion earlier in this chapter. For example, the d distances of two dots—beginning with the dot at the top of the circle and skipping one dot each time—have corresponding c distances of 5, 5, 5, 5, and 4 crossing lines respectively. These distances are depicted in the second part of the figure as an illustration. Similarly, all of the other d distances correspond to c distances of two consecutive numbers, as recorded in the table.

Figure 1.9 Checking a five-dot circle diagram to see if it is maximally even according to the definition

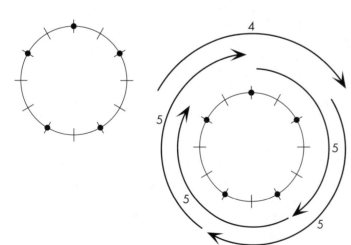

CLOCKWISE DISTANCE BETWEEN DOTS	
d distance	**c distance**
1	2, 3
2	4, 5
3	7, 8
4	9, 10

Testing Other Diagrams for Maximal Evenness

Exercise 1.6 provides an opportunity to check each of the musical structures we previously identified as spread out as much as possible to determine if they are indeed maximally even according to our definition. Each of these musical structures (except the diminished seventh chord and pentatonic scale that were solved for you in Figures 1.7 and 1.9, respectively) is given in staff notation. For each structure, first plot the notes on the circle diagrams, then complete the table. Finally, based on the table, determine if the musical structure examined is maximally even according to the definition. Chromatic notes are labeled in these circle diagrams with sharps only, for consistency; to plot a flat note, use the corresponding enharmonically equivalent note (for example, A♯ for B♭).

EXERCISE 1.6 Check the musical structures indicated on the staves to determine if they are maximally even according to the definition. Plot each structure on the circle diagram, and complete each interval table.

a.

CLOCKWISE DISTANCE BETWEEN DOTS	
d distance	**c distance**

Maximally even? _____

b.

CLOCKWISE DISTANCE BETWEEN DOTS	
d distance	**c distance**

Maximally even? _____

c.

CLOCKWISE DISTANCE BETWEEN DOTS	
d distance	**c distance**

Maximally even? _____

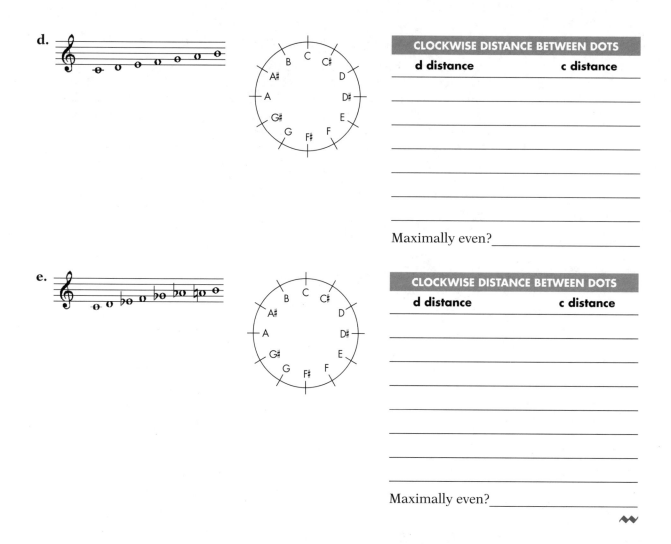

The answers for this exercise are given in Solution 1.6. Of course, each of the structures is maximally even according to the formal definition, as you can see from the tables you completed in the exercise. Because this exercise is a straightforward problem of plotting notes and counting lines around circles, we shall discuss only one of the musical structures, the familiar diatonic collection (appearing as a C major scale in Exercise 1.6), in further detail. For the other musical structures, you can simply check your answers against those provided.

Intervals and Distances in Seven-Dot Circles

The seven-dot circle diagram, shown in Solution 1.6d, represents a special case that needs additional explanation because it corresponds to such a familiar musical construct—the major scale. In the table shown in Solution 1.6d, each d distance represents the interval types (or sizes) of seconds, thirds, fourths, fifths, sixths, and sevenths, in the traditional sense. For example, each d distance of 1 (between adjacent dots) is equivalent to the interval type of a second, each d distance of 2 (skipping one dot) is equivalent to the interval type of a third, each d distance of 3 (skipping two dots) is equivalent to the interval type of a fourth, and so on. This correspondence between interval types and d distances is applicable to seven-note collections, because the traditional interval identification system was originally devised for seven-note collections (as represented by the lines and spaces of the musical staff). The table shows that for each of these interval types (or d distances), there are only two interval qualities (or c distances), and the intervals have consecutive numbers of half steps. Therefore, because the major scale is maximally even, all seconds in the major scale must be either major or minor (with 2 half steps or 1 half step, respectively), all thirds must be either major or minor (with 4 or 3 half steps), all fourths must be either augmented or perfect (6 or 5 half steps), all fifths must be either perfect or diminished (7 or 6 half steps), all sixths must be either major or minor (9 or 8 half steps), and all sevenths must be either major or minor (11 or 10 half steps). This list presents all of the primary intervals taught in introductory music theory courses. The formal definition of maximal evenness from a musical perspective, applied to a seven-note scale, rests firmly upon the knowledge of these familiar intervals.

SOLUTION 1.6 Circle diagrams and the corresponding interval tables for maximally even structures of two, three, six, seven, and eight notes

a.

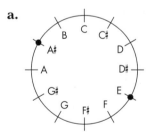

CLOCKWISE DISTANCE BETWEEN DOTS	
d distance	**c distance**
1	6
Maximally even?	yes

b.

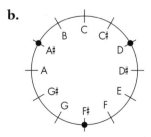

CLOCKWISE DISTANCE BETWEEN DOTS	
d distance	**c distance**
1	4
2	8
Maximally even?	yes

c.

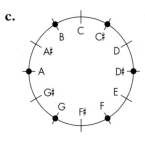

CLOCKWISE DISTANCE BETWEEN DOTS	
d distance	**c distance**
1	2
2	4
3	6
4	8
5	10
Maximally even?	yes

d.

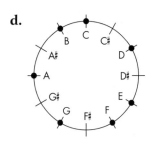

CLOCKWISE DISTANCE BETWEEN DOTS	
d distance	**c distance**
1	1, 2
2	3, 4
3	5, 6
4	6, 7
5	8, 9
6	10, 11
Maximally even?	yes

e.

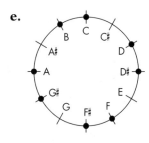

CLOCKWISE DISTANCE BETWEEN DOTS	
d distance	**c distance**
1	1, 2
2	3
3	4, 5
4	6
5	7,8
6	9
7	10, 11
Maximally even?	yes

Testing Other Scales for Maximal Evenness

We will now examine a few other familiar scales to determine if they are maximally even. In Exercise 1.7, plot the indicated scales on the circle diagrams. For each scale, complete the given table to determine whether or not the scale is maximally even. Play each of these scales on a piano and listen to the interval patterns formed; try to hear these scales as maximally even or uneven, based on your findings in the exercise.

EXERCISE 1.7 Plot the indicated scales on the circle diagrams. For each scale, complete the given table to determine if the scale is maximally even.

a. E harmonic minor scale

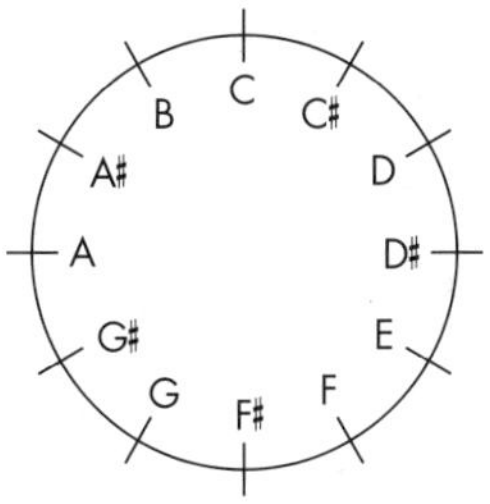

CLOCKWISE DISTANCE BETWEEN DOTS	
d distance	**c distance**

Maximally even?_____

b. B melodic minor scale (ascending)

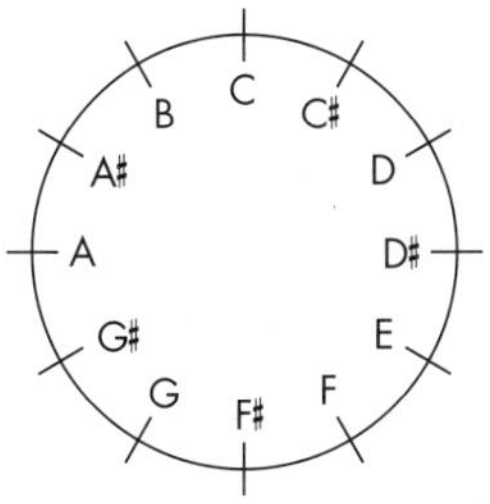

CLOCKWISE DISTANCE BETWEEN DOTS	
d distance	**c distance**

Maximally even?_____

c. F♯ melodic minor scale (descending)

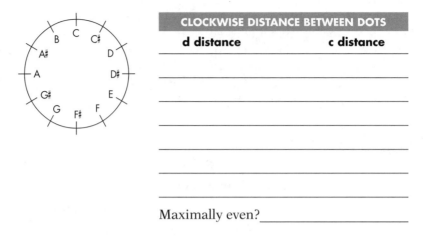

CLOCKWISE DISTANCE BETWEEN DOTS	
d distance	**c distance**

Maximally even? _____

Descending Melodic Minor

As you probably have discovered, and as shown in Solution 1.7, only one of the scales in Exercise 1.7 is maximally even. Why do you suppose the F♯ descending melodic minor scale, depicted in Solution 1.7c, is maximally even? Compare this scale with the other seven-note scales we discussed earlier in the chapter (shown in Figure 1.5 on p. 20). Are there any similarities? The descending melodic minor scale (that is, any descending melodic minor scale, not just the one built on F♯) has exactly the same series of intervals as the natural minor scale—which we already identified as maximally even in Figure 1.5, by visual inspection rather than by using the formal definition. The other two scales in the exercise and Solution 1.7 have different intervals, as determined in reference to the tables and in the discussion that follows.

Ascending Melodic Minor

Continuing with our discussion of the scales in Exercise 1.7 in reverse order, the ascending melodic minor scale (again, any ascending melodic minor scale, not just the B scale represented in Solution 1.7b) is perhaps closest to maximally even; only the d distances of 3 and 4 have three different c distances, rather than the required one or two consecutive c distances. In musical terms, the scale includes diminished fourths (c distance of 4: A♯–D), perfect fourths (c distance of 5: B–E, C♯–F♯, F♯–B, G♯–C♯), and augmented fourths (c distance of 6: D–G♯, E–A♯)—each with a d distance of 3. From another perspective, with the same pairs of notes in the opposite order and with the diminished and augmented intervals exchanged (due to inversion), the scale contains diminished fifths (c distance of 6: G♯–D, A♯–E), perfect fifths (c distance of 7: E–B, F♯–C♯, B–F♯, C♯–G♯), and augmented fifths (c distance of 8: D–A♯)—each with a d distance of 4. Unlike these fourths and fifths, the other d distances (1, 2, 5, and 6) conform to the definition, having only two consecutive c distances for each d distance. But because some d distances have more than two c distances, the ascending melodic minor scale is not maximally even. Interestingly, an alternative definition that has been developed for maximal evenness, but which we will not explore further in this text, uses an approach that calculates precise values for the relative weights of musical structures in terms of their evenness to determine that the ascending melodic minor scale is the *second-most* maximally even seven-note structure.[6]

Harmonic Minor

Finally, the harmonic minor scale, represented in Solution 1.7a, regardless of tonic note, is not maximally even.[7] In this scale, many d distances have 3 different c distances. A closer inspection of the d distance of 1 dot in the table—or the interval of a second, in musical terms—reveals an essential aspect of the harmonic minor scale, one that seems to have plagued composers of tonal music for centuries. The fact that this scale has three different seconds—minor (c = 1), major (c = 2), and augmented (c = 3)—can cause scalar passages to sound disjointed. On the other hand, maximally even scales, plus the *almost*-maximally even ascending melodic minor scale, have only two sizes between adjacent scale steps, providing smoother scalar passages. Perhaps an analogy will help illustrate the importance of this observation.

Circle diagrams and the corresponding interval tables for different forms of the minor scale

a. E harmonic minor scale

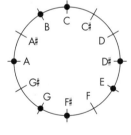

CLOCKWISE DISTANCE BETWEEN DOTS	
d distance	c distance
1	1, 2, 3
2	3, 4
3	4, 5, 6
4	6, 7, 8
5	8, 9
6	9, 10, 11
Maximally even?	no

b. B melodic minor scale (ascending)

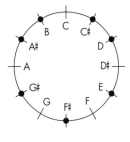

CLOCKWISE DISTANCE BETWEEN DOTS	
d distance	**c distance**
1	1, 2
2	3, 4
3	4, 5, 6
4	6, 7, 8
5	8, 9
6	10, 11

Maximally even?_____no_____

c. F♯ melodic minor scale (descending)

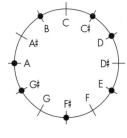

CLOCKWISE DISTANCE BETWEEN DOTS	
d distance	**c distance**
1	1, 2
2	3, 4
3	5, 6
4	6, 7
5	8, 9
6	10, 11

Maximally even?_____yes_____

The Stepping-Stone Analogy

Imagine crossing a stream by stepping from stone to stone. The task would be easier if the distance between one stone and the next was the same for all stones, rather than having different distances to gauge for each of your steps. With the stones placed in a maximally even way, you would become accustomed to the distances required to step from stone to stone, even if there were two different distances. However, crossing the stream would be more complicated if most of the stones were fairly evenly spaced, except for one larger distance between two stones, requiring a short leap to avoid getting wet. You might become used to the smaller distances between stones and increase your speed—only to discover (too late) that one of the stones is not where you expected it (and get soaked!).

Traversing the harmonic minor scale in a musical setting presents the same kind of problem, where all of the steps are either half steps or whole steps, except one augmented second that seems to appear out of nowhere. Hence, composers usually avoided the spot where the augmented second occurs, instead moving stepwise below or above it, but not usually crossing that precarious distance in a stepwise manner. The fact that the harmonic minor scale is not maximally even, due to the three different qualities of each interval type, reflects this important facet of the harmonic minor scale—an issue that will return again and again throughout your study of tonal music.

SUMMARY AND EXTENSIONS

Maximal Evenness and the Black and White Keys

The distinct spatial patterns illustrated in the puzzles with which we began this chapter all correspond to identifiable musical structures. Placing dots around a circle of twelve crossing lines (representing the twelve notes in an octave) so that the dots are spread out as much as possible in *every* case generates musically significant counterparts. Moreover, the maximally even seven-dot circle corresponds to perhaps the most familiar collection of notes employed by composers of Western music—the diatonic collection, which encompasses the major, natural minor, and descending melodic minor scales, as well as the modes. Maximally even structures are fundamental to our understanding of music, extending even to the distribution of notes and basic layout of the piano keyboard. Both the white keys, corresponding to the diatonic collection, and the black keys, forming the pentatonic scale, are maximally even. This fundamental arrangement of black and white keys, and therefore the intervals between the notes they represent, continues to represent our standard approach to the organization of notes, prevailing even in most computer applications through the MIDI standard.

The complementary relationship between the seven-note and five-note maximally even structures, so familiar because of their arrangement on the keyboard, also can be seen in the circle diagrams. Because both the pattern of dots and the pattern of spaces in these diagrams are maximally even, the pentatonic collection and the diatonic collection are complements of each other. Likewise, the diminished seventh chord and the octatonic scale are complementary. Strangely enough, the complement of the whole-tone scale is another whole-tone scale, as you can see easily by comparing the arrangement of dots in any of the six-dot circle diagrams with the arrangement of crossing lines that lack dots.[8]

Maximally Even Structures in Context

As you progress in your study of music theory, you will encounter musical contexts for many of the maximally even structures we have worked with in this book. Some of these structures—such as the augmented triad, the pentatonic scale, the whole-tone scale, and the octatonic scale—began to be used extensively by early twentieth-century composers such as Claude Debussy (1862–1918), Aleksandr Scriabin (1872–1915), and many others. Other maximally even structures—tritones, diminished seventh chords, and diatonic scales—are fundamental to music of the eighteenth and nineteenth centuries (and much music of the twentieth century as well) and will be essential components of your continuing study of tonal music theory. The tritone is indispensable to tonal structure in terms of the resolution of dissonance; the diminished seventh chord becomes an increasingly common means of emphasizing certain notes, chords, and keys; and the diatonic collection provides the primary scalar material explored by many composers working in a tonal tradition.

Interval Inversion

The uniformity of the diatonic collection, evident in its maximally even structure, may be viewed from another perspective that involves counting intervals around a circle. This view relies on an understanding of interval inversion, mentioned earlier in connection with the tables we constructed to identify maximal evenness formally. In musical terms, inverting an interval involves reversing the perspective between two notes in relation to each other so that the lower of the two notes becomes the higher of the two notes, and vice versa. For example, the interval from C up to E♭, inverted, becomes the interval from E♭ up to C. In traditional intervallic terms, inverted interval types (or sizes) sum to nine, and inverted interval qualities switch between major and minor, augmented and diminished, or perfect and perfect. Thus, a *minor third*, from C up to E♭, becomes a *major sixth* (summing to nine), from E♭ up to C. Likewise, a diminished third becomes an augmented sixth, and a perfect fifth becomes a perfect fourth. However, for the remainder of this chapter, we will refer to these inversionally related pairs using c distances.

In terms of c distances (or half steps), inverted intervals sum to twelve (the number of half steps in an octave). Thus, a c distance of 3, inverted, becomes a c distance of 9 (as illustrated in Figure 1.10a), and a c distance of 6, inverted, remains a c distance of 6 (as shown in Figure 1.10b)—each pair of intervals summing to twelve. All inversionally related pairs of intervals behave in this way; therefore, inversions essentially involve two ways of determining the distance between a *single* pair of dots on a circle (or a single pair of notes). Consequently, if we wish to count the number of times that each interval occurs in a circle diagram, we can simply ignore the larger c distance for each pair of notes (effectively counting only c distances from 1 to 6).[9] In this way we ensure that inversionally related interval pairs will be counted only once. For example, in Figure 1.10a, because there is only one pair of dots, we count only one c distance (3). Special care must be taken with c distances of 6, because the inversion of 6 is 6. Therefore, in Figure 1.10b, as there is only one pair of dots, we count only one c distance (6).

Figure 1.10 Circle diagrams showing that the inversion of a c distance of 3 is a c distance of 9, and the inversion of a c distance of 6 is a c distance of 6. Inversionally related c distances sum to twelve (the number of half steps in an octave).

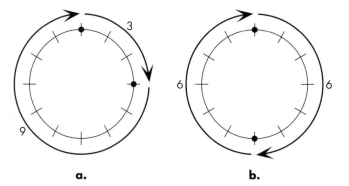

a.   b.

Interval Content and the Deep Scale Property

If we tabulate the number of times each c distance appears in a circle diagram (taking only the smaller of each inversional pair to avoid double-counting), we will be able to observe another special property of the diatonic collection. As shown in the first circle diagram of Figure 1.11, the c distance of 1 appears two times in the circle corresponding to the diatonic collection (in this case, the F♯ descending melodic minor scale taken from Exercise 1.7c). Likewise, in the remaining circle diagrams in the figure, and indicated by the brackets, the c distance of 2 appears five times, 3 appears four times, 4 appears three times, 5 appears six times, and 6 appears one time. The bottom of the figure displays these c distances in a table that shows the number of times that each c distance appears in the diatonic collection (excluding the larger intervals of inversionally related pairs).[10] It is easy to see in the table that each c distance appears a different number of times (a different value appears in each of the lower boxes in the table). This special property of the diatonic collection (where each interval appears a different number of times) is termed the *deep scale property*.[11]

Figure 1.11 Tabulating c distances (for the smaller of inversionally related pairs) in the diatonic collection

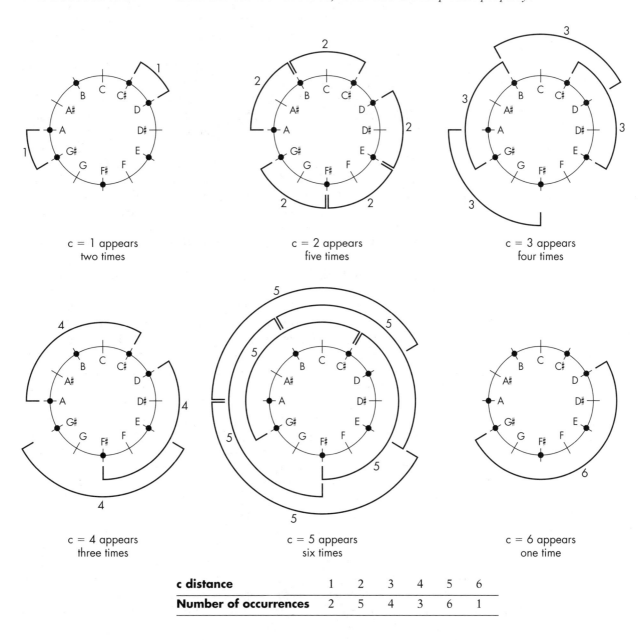

c = 1 appears two times

c = 2 appears five times

c = 3 appears four times

c = 4 appears three times

c = 5 appears six times

c = 6 appears one time

c distance	1	2	3	4	5	6
Number of occurrences	2	5	4	3	6	1

Modulation, Common Tones, and the Deep Scale

One of the most musically significant aspects of the deep scale property is its relationship to modulation and common tones. *Modulation* refers to the process through which keys change in a tonal composition. The table of c distance occurrences for the diatonic collection (bottom of Figure 1.11) indicates the number of notes that will be held in common (called *common tones*) if the diatonic collection is transposed by each particular c distance, with one exception (to be discussed later). And the deep scale property (where each interval appears a different number of times) ensures that (again with one exception) there will be a different number of common tones associated with each transposition level.[12]

This direct relationship between the number of occurrences of each c distance in the diatonic collection and the number of common tones between two transposed diatonic collections can be seen easily by comparing the key signatures of various diatonic collections.[13] For example, because the c distance of 1 appears two times in the diatonic collection, as shown in the table at the bottom of Figure 1.11, a diatonic collection that is transposed by a c distance of 1 will have two common tones. Thus, the key of C major (with no flats or sharps) transposed up a c distance of 1 yields the key of D♭ major (with five flats). Because five of the notes in the C major collection are now flat (the five flats in the D♭ major key signature), the two remaining notes (of the seven total diatonic notes) must be common tones (C and F), as indicated in the table. Likewise, because the c distance of 2 appears five times in the table, a diatonic collection that is transposed by a c distance of 2 will have five common tones. Thus, the key of C major transposed up a c distance of 2 yields the key of D major (with two sharps). Because two of the notes in the original collection are now sharp (C♯ and F♯, as depicted in the key signature for D major), five notes are held in common between the two collections (D, E, G, A, and B).

In the same way, keys that are transpositionally related by c distances of 3 (such as C major and E♭ major) have four common tones, c distances of 4 (such as C major and E major) have three common tones, and c distances of 5 (such as C major and F major) have six common tones—as revealed by the table and as suggested by comparing the respective pairs of key signatures. The one exception involves the c distance of 6, where the enharmonic spelling of one of the notes yields one more common tone than the table indicates. Thus, C major and F♯ major (a c distance of 6) have two common tones, though their key signatures are six sharps apart. Here, the E♯ included in the F♯ major collection also appears in the C major collection, enharmonically spelled as F♮. Hence, there are two common tones, rather than the one (B) suggested by the table. But this inconsistency may be accounted for by remembering that we initially counted c distances of 6 only once in constructing the original table, as shown in Figure 1.11 and as discussed earlier, to avoid duplication of enharmonically equivalent intervals. The "extra" enharmonically equivalent common tone is related to this "extra" enharmonically equivalent c distance of 6.

The significance of these findings for modulation is that the closeness of the relationship between any two keys, in terms of the number of common tones, (with one exception) is uniquely associated with the transpositional proximity of the keys (or c distance apart). Thus, any pair of keys that is related by a c distance of 5 (with six common tones) will always have a much closer relationship than a pair of keys that is related by a c distance of 1 (with only two common tones), despite the adjacent location of the latter two keys. And in the tonal-music literature, modulations to keys that are a perfect fifth or fourth apart (a c distance of 5) occur more frequently and are considered to be *closely* (or *nearly*) *related*, whereas keys that border each other chromatically (c distance of 1) are viewed as *distantly related* (or *foreign*).

You may wish to examine on your own the other collections from this chapter—and perhaps some collections of your own design—to see if they have the deep scale property as well, following the procedures outlined earlier in this section. We have now observed two special properties of the diatonic collection in this chapter: The diatonic collection is maximally even, and it is also a deep scale.

A Look Forward

Along with this brief overview of the deep scale property (a natural extension of some of the ideas presented earlier about intervals and circle diagrams), this chapter has provided a thorough introduction to maximal evenness, has led you to identify musical structures that are maximally even, and has provided an abstract, non-staff-based structure in which to consider musical concepts. You have worked with intervals by counting half steps (c distances) and relative distances between notes (d distances), and you have compared the various forms of the minor scale in some detail. You have also examined augmented triads; diminished seventh chords; and whole-tone, pentatonic, and octatonic scales in abstract ways—laying a foundation for more advanced study in the future. In the next chapter we will continue to explore all of these collections, but we will begin with the more familiar diatonic.

INTERVAL PATTERNS AND MUSICAL STRUCTURES | 2

DIATONIC INTERVAL PATTERNS

In Chapter 1 we observed that the diatonic collection is the only seven-note collection that is maximally even in a twelve-note chromatic universe. (I use the term *chromatic universe* to indicate all available notes in a given system. In the usual twelve-note system, chromatic universe refers to the unordered twelve notes of the chromatic scale.) We will return to the idea of maximal evenness in Chapter 3. In this chapter we explore other important properties of the diatonic collection. Again we take a self-discovery approach: Rather than beginning with an explanation of the property at work, this chapter presents exercises and discussion designed to help you reach your own conclusions about the patterns you observe as you solve the exercises. Through this approach, we postpone defining the properties of the diatonic collection that we will explore, as in the first chapter, thus giving you an opportunity to develop your own definitions.

Interval Identification in Transposed Series

We begin with a series of simple interval identification problems.[1] Solving these problems will allow you to observe an important property of the diatonic collection, and these exercises also allow you to practice identifying intervals. Exercise 2.1 contains several groups of three-note series, all in diatonic contexts. Transpose each of these three-note series diatonically so that a series begins on each note of the diatonic collection in each group. That is, transpose the series within a single diatonic collection, moving the whole series up by the interval of a second each time, without adjusting the key signature. Next, identify the intervals between every pair of adjacent notes for each version of the series; include the interval from the last note in the series up to the octave above the first note. For each interval identification, give both the number of half steps and the interval name by type (or size) and quality. The total number of half steps in each interval pattern should sum to twelve. Finally, compare the intervals you have identified, in order, for every three-note series in each group. For each group of three-note series, determine how many distinct interval patterns appear in the group.

Three-Note Series

Exercise 2.1 presents three-note series in various diatonic contexts. Exercise 2.1a traces a stepwise pattern in C major, Exercise 2.1b involves a diatonic step followed by a skip of a third in C major, Exercise 2.1c delineates a step and then a leap of a fourth in F major, and Exercise 2.1d uses a skip and then a step in E♭ major. Implicit in each pattern is a return to the initial note to complete the octave. This return to the starting place is shown in parentheses for the first two series in Exercise 2.1a but is assumed elsewhere. You can include these parenthetical notes in your solutions if you wish, but you must include the *interval* required to *complete* the octave in each of your interval patterns.

Diatonic Transposition and Circle Diagrams

Comparing these diatonic transposition exercises to the circle diagrams, we are keeping a constant pattern of d distances as we rotate (or transpose diatonically) around the circle while observing the c distances between the selected dots in the resulting patterns (shown by half steps as well as by interval quality and type). The intervals that complete the octave in these series of notes correspond to the c distances necessary to complete the circle by returning to the initial dot.

Counting Distinct Interval Patterns

The first three-note series in Exercise 2.1a, shown as a sample, has an interval pattern of 2–2–8 or M2–M2–m6, and the second has an interval pattern of 2–1–9 or M2–m2–M6. The last interval of each of these patterns is the interval required to complete the octave (shown in parentheses). You may omit the parenthetical *note* if you wish, but you must include this *interval*. Identify the other five series of notes in the same way. Some of the patterns will appear more than once. Although you need to label all of the interval patterns formed (even if they are repetitions of earlier patterns), you need to count and indicate at the end of each group only the number of *distinct* (or different) interval patterns that appear in each group. Also, it is important that you write out all of the interval patterns, not just the distinct ones, because you will use these results again later in this chapter.

For each group, transpose the series diatonically so that a series begins on each note of the corresponding major scale. Identify the intervals between each pair of adjacent notes as indicated. For each group, determine how many distinct interval patterns appear.

a. C major

b. C major

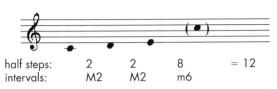

half steps: 2   2   8   = 12
intervals: M2   M2   m6

half steps:
intervals:

half steps: 2   1   9   = 12
intervals: M2   m2   M6

half steps:
intervals:

(etc.)

half steps: = 12
intervals:

half steps:
intervals:

half steps:
intervals:

half steps:
intervals:

half steps:
intervals:

half steps:
intervals:

half steps:
intervals:

half steps:
intervals:

half steps:
intervals:

half steps:
intervals:

How many distinct interval patterns appear in the group? _____

How many distinct interval patterns appear in the group? _____

c. F major

half steps:
intervals:

half steps:
intervals:

half steps:
intervals:

half steps:
intervals:

half steps:
intervals:

half steps:
intervals:

half steps:
intervals:

How many distinct interval patterns appear in the group? _____

d. E♭ major

half steps:
intervals:

half steps:
intervals:

half steps:
intervals:

half steps:
intervals:

half steps:
intervals:

half steps:
intervals:

half steps:
intervals:

How many distinct interval patterns appear in the group? _____

Four-Note Series

Before we discuss the solutions to Exercise 2.1, we will examine several groups of four-note series in Exercise 2.2 in the same way. As in the previous exercise, diatonically transpose each series, and identify the intervals by half steps and type and quality. Again, each group should have a total of seven series, and the number of half steps in each pattern of intervals should sum to twelve because the patterns assume the completion of the octave. (The notes needed to complete the octave are not shown in parentheses in the initial patterns, but they are still an essential implied part of the exercise.) Finally, determine how many distinct interval patterns appear in each group. (You should complete the exercise on the following two pages before continuing in the text.) After you complete Exercises 2.1 and 2.2, it would be beneficial to play all of these series on a piano or some other instrument in order to experience these intervals aurally. Listen carefully to the sounds these series make. Compare the series that have identical interval patterns as well as those that are distinct within each group. Try to *hear* what makes these interval patterns sound similar to or distinct from one another.

Comparing Interval Patterns

The answers for these two exercises are provided in Solutions 2.1 and 2.2. Carefully check your transposed series, and especially your interval identifications, to be sure that you have correctly completed the exercises. Observe that the note required to complete the octave in each series is assumed in each interval pattern, but not shown (except for the first two samples). If you desired, you may have included these parenthetical notes (an octave above each initial note) in your solutions as discussed earlier. Exercise 2.1a has three different interval patterns: 2–2–8 (or M2–M2–m6), 2–1–9 (M2–m2–M6), and 1–2–9 (m2–M2–M6). Exercise 2.1b has three different interval patterns: 2–3–7 (M2–m3–P5), 1–4–7 (m2–M3–P5), and 2–4–6 (M2–M3–d5). Similarly, Exercises 2.1c and 2.1d each have three different interval patterns: 2–5–5 (M2–P4–P4), 1–6–5 (m2–A4–P4), and 1–5–6 (m2–P4–A4) for Exercise 2.1c; and 4–1–7 (M3–m2–P5), 3–2–7 (m3–M2–P5), and 4–2–6 (M3–M2–d5) for Exercise 2.1d. On the other hand, Exercise 2.2a has four different interval patterns: 2–2–1–7, 2–1–2–7, 1–2–2–7, and 2–2–2–6, shown here in half steps for simplicity. Likewise, Exercise 2.2b has four different interval patterns: 2–2–3–5, 2–1–4–5, 1–2–4–5, and 1–2–3–6. Finally, Exercises 2.2c and 2.2d each have four different interval patterns: 2–3–2–5, 1–4–2–5, 2–4–1–5, and 1–4–1–6 for Exercise 2.2c; and 4–1–4–3, 3–2–4–3, 3–2–3–4, and 4–2–3–3 for Exercise 2.2d.

For each group, transpose the series diatonically so that a series begins on each note of the corresponding major scale. Identify the intervals between each pair of adjacent notes as indicated. For each group, determine how many distinct interval patterns appear.

a. C major

half steps:
intervals:

half steps:
intervals:

half steps:
intervals:

half steps:
intervals:

half steps:
intervals:

half steps:
intervals:

half steps:
intervals:

How many distinct interval patterns appear in the group? _____

b. D major

half steps:
intervals:

half steps:
intervals:

half steps:
intervals:

half steps:
intervals:

half steps:
intervals:

half steps:
intervals:

half steps:
intervals:

How many distinct interval patterns appear in the group? _____

c. F major

half steps:
intervals:

half steps:
intervals:

half steps:
intervals:

half steps:
intervals:

half steps:
intervals:

half steps:
intervals:

half steps:
intervals:

How many distinct interval patterns appear in the group? _____

d. B♭ major

half steps:
intervals:

half steps:
intervals:

half steps:
intervals:

half steps:
intervals:

half steps:
intervals:

half steps:
intervals:

half steps:
intervals:

How many distinct interval patterns appear in the group? _____

a. C major

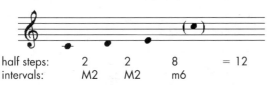

half steps:	2	2	8	= 12
intervals:	M2	M2	m6	

half steps:	2	1	9	= 12
intervals:	M2	m2	M6	

half steps:	1	2	9	= 12
intervals:	m2	M2	M6	

half steps:	2	2	8
intervals:	M2	M2	m6

half steps:	2	2	8
intervals:	M2	M2	m6

half steps:	2	1	9
intervals:	M2	m2	M6

half steps:	1	2	9
intervals:	m2	M2	M6

How many distinct interval patterns appear in the group? _____3_____

b. C major

half steps:	2	3	7
intervals:	M2	m3	P5

half steps:	2	3	7
intervals:	M2	m3	P5

half steps:	1	4	7
intervals:	m2	M3	P5

half steps:	2	4	6
intervals:	M2	M3	d5

half steps:	2	3	7
intervals:	M2	m3	P5

half steps:	2	3	7
intervals:	M2	m3	P5

half steps:	1	4	7
intervals:	m2	M3	P5

How many distinct interval patterns appear in the group? _____3_____

c. F major

half steps: 2   5   5
intervals: M2   P4   P4

half steps: 2   5   5
intervals: M2   P4   P4

half steps: 1   6   5
intervals: m2   A4   P4

half steps: 2   5   5
intervals: M2   P4   P4

half steps: 2   5   5
intervals: M2   P4   P4

half steps: 2   5   5
intervals: M2   P4   P4

half steps: 1   5   6
intervals: m2   P4   A4

How many distinct interval patterns appear in the group? ___3___

d. E♭ major

half steps: 4   1   7
intervals: M3   m2   P5

half steps: 3   2   7
intervals: m3   M2   P5

half steps: 3   2   7
intervals: m3   M2   P5

half steps: 4   2   6
intervals: M3   M2   d5

half steps: 4   1   7
intervals: M3   m2   P5

half steps: 3   2   7
intervals: m3   M2   P5

half steps: 3   2   7
intervals: m3   M2   P5

How many distinct interval patterns appear in the group? ___3___

Transposed series of notes drawn from major scales and the resulting interval patterns

a. C major

half steps:	2	2	1	7
intervals:	M2	M2	m2	P5

half steps:	2	1	2	7
intervals:	M2	m2	M2	P5

half steps:	1	2	2	7
intervals:	m2	M2	M2	P5

half steps:	2	2	2	6
intervals:	M2	M2	M2	d5

half steps:	2	2	1	7
intervals:	M2	M2	m2	P5

half steps:	2	1	2	7
intervals:	M2	m2	M2	P5

half steps:	1	2	2	7
intervals:	m2	M2	M2	P5

How many distinct interval patterns appear in the group? ___4___

b. D major

half steps:	2	2	3	5
intervals:	M2	M2	m3	P4

half steps:	2	1	4	5
intervals:	M2	m2	M3	P4

half steps:	1	2	4	5
intervals:	m2	M2	M3	P4

half steps:	2	2	3	5
intervals:	M2	M2	m3	P4

half steps:	2	2	3	5
intervals:	M2	M2	m3	P4

half steps:	2	1	4	5
intervals:	M2	m2	M3	P4

half steps:	1	2	3	6
intervals:	m2	M2	m3	A4

How many distinct interval patterns appear in the group? ___4___

c. F major

half steps: 2   3   2   5
intervals: M2   m3   M2   P4

half steps: 2   3   2   5
intervals: M2   m3   M2   P4

half steps: 1   4   2   5
intervals: m2   M3   M2   P4

half steps: 2   4   1   5
intervals: M2   M3   m2   P4

half steps: 2   3   2   5
intervals: M2   m3   M2   P4

half steps: 2   3   2   5
intervals: M2   m3   M2   P4

half steps: 1   4   1   6
intervals: m2   M3   m2   A4

How many distinct interval patterns appear in the group? ____4____

d. B♭ major

half steps: 4   1   4   3
intervals: M3   m2   M3   m3

half steps: 3   2   4   3
intervals: m3   M2   M3   m3

half steps: 3   2   3   4
intervals: m3   M2   m3   M3

half steps: 4   2   3   3
intervals: M3   M2   m3   m3

half steps: 4   1   4   3
intervals: M3   m2   M3   m3

half steps: 3   2   3   4
intervals: m3   M2   m3   M3

half steps: 3   2   3   4
intervals: m3   M2   m3   M3

How many distinct interval patterns appear in the group? ____4____

Forming a Hypothesis

After you have completed and checked these interval identification exercises, return to Exercises 2.1 and 2.2 and examine each of the transposed groups in terms of the number of distinct interval patterns formed for each group and the number of notes in each group. In Exercise 2.3, make a generalized statement, or hypothesis, to explain your observations. In whatever words seem appropriate to you, explain what you are observing in the series or what general principle unites the interval patterns in each group. You may not have enough information to verify your assertion, and you may not be able to explain why these patterns work as they do. Nevertheless, make your best effort in proposing a generalized statement before turning to the discussion of the solution that follows.

EXERCISE 2.3

Based on your examination of the number of notes in a series and the number of different interval patterns formed for each group in Exercises 2.1 and 2.2, make a generalized statement to explain your observations.

_____

_____

_____

_____

Based on your observations, you may have created a statement somewhat similar to the answer provided in Solution 2.3 (on p. 59). The exact wording of this solution is not important; in fact, eventually we will see that an elegant three-word phrase captures the answer suggested here. However, before we discuss this formal designation, we will continue with our self-discovery approach and explore our own informal ideas about the issue. Regardless of how you expressed your statement, your response should acknowledge that the number of notes in a series matches the number of interval patterns formed.

Five-Note and Six-Note Series

Before we can be satisfied with our hypothesis, however, we must see if this generalized statement holds true for other sized series of notes—not merely patterns formed by series of three and four notes alone. Exercise 2.4 contains a couple of five-note and six-note series to be investigated. For each given series, diatonically transpose the series to produce a group of seven related series. Determine the interval patterns formed by each series, and tabulate the number of distinct interval patterns that appear. Again, you can show the parenthetical notes required to complete octaves if you desire, though these notes are not shown in these solutions; however, you must include the intervals corresponding to these octave completions in the interval patterns.

For each group, transpose the series diatonically so that a series begins on each note of the corresponding major scale. Identify the intervals between each pair of adjacent notes as indicated. For each group, determine how many distinct interval patterns appear.

a. C major

half steps:
intervals:

half steps:
intervals:

half steps:
intervals:

half steps:
intervals:

half steps:
intervals:

half steps:
intervals:

half steps:
intervals:

How many distinct interval patterns appear in the group? _____

b. D♭ major

half steps:
intervals:

half steps:
intervals:

half steps:
intervals:

half steps:
intervals:

half steps:
intervals:

half steps:
intervals:

half steps:
intervals:

How many distinct interval patterns appear in the group? _____

c. C major

half steps:
intervals:

half steps:
intervals:

half steps:
intervals:

half steps:
intervals:

half steps:
intervals:

half steps:
intervals:

half steps:
intervals:

How many distinct interval patterns appear in the group? _____

d. E major

half steps:
intervals:

half steps:
intervals:

half steps:
intervals:

half steps:
intervals:

half steps:
intervals:

half steps:
intervals:

half steps:
intervals:

How many distinct interval patterns appear in the group? _____

A generalized statement to describe the relationship between the number of notes in a series and the number of different interval patterns formed for each group in Exercises 2.1 and 2.2

The number of notes in a series equals the number of different

interval patterns that can be formed by transposing that series

diatonically.

Solution 2.4 provides the answers for these problems. As before, carefully check your transposed series—especially your interval identifications—to be sure that you have completed the exercises correctly. As you likely have concluded, assuming your interval identifications are correct, the number of notes in a series of five or six notes also precisely indicates the number of different interval patterns formed. Five-note series yield five different interval patterns, and six-note series produce six different interval patterns. Although the series have become longer and the patterns of intervals perhaps more complicated, our hypothesis holds true for these series as well. If you are skeptical, you might wish to check all of the other series of notes to be sure that the general statement applies in all cases. It would be prudent to do such an exhaustive inquiry, but to save time we will limit ourselves to an examination of all possible diatonic two-note series.

Two-Note Series

In Exercise 2.5, construct all six possible groups of two-note series in the diatonic collection corresponding to the C major scale. Determine the interval pattern formed by each series (including the interval required to complete the octave), and for each group determine the number of different interval patterns formed. We expect to find—as was consistent with three-, four-, five-, and six-note series—exactly two distinct interval patterns for each group.

Transposed series of notes drawn from major scales and the resulting interval patterns

a. C major

half steps: 2   2   1   2   5
intervals: M2   M2   m2   M2   P4

half steps: 2   1   2   2   5
intervals: M2   m2   M2   M2   P4

half steps: 1   2   2   2   5
intervals: m2   M2   M2   M2   P4

half steps: 2   2   2   1   5
intervals: M2   M2   M2   m2   P4

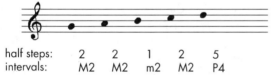

half steps: 2   2   1   2   5
intervals: M2   M2   m2   M2   P4

half steps: 2   1   2   2   5
intervals: M2   m2   M2   M2   P4

half steps: 1   2   2   1   6
intervals: m2   M2   M2   m2   A4

How many distinct interval patterns appear in the group? ____5____

b. D♭ Major

half steps: 2   2   3   4   1
intervals: M2   M2   m3   M3   m2

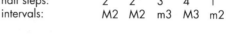

half steps: 2   1   4   3   2
intervals: M2   m2   M3   m3   M2

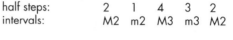

half steps: 1   2   4   3   2
intervals: m2   M2   M3   m3   M2

half steps: 2   2   3   4   1
intervals: M2   M2   m3   M3   m2

half steps: 2   2   3   3   2
intervals: M2   M2   m3   m3   M2

half steps: 2   1   4   3   2
intervals: M2   m2   M3   m3   M2

half steps: 1   2   3   4   2
intervals: m2   M2   m3   M3   M2

How many distinct interval patterns appear in the group? ____5____

c. C major

half steps: 2   2   1   2   2   3
intervals: M2   M2   m2   M2   M2   m3

half steps: 2   1   2   2   2   3
intervals: M2   m2   M2   M2   M2   m3

half steps: 1   2   2   2   1   4
intervals: m2   M2   M2   M2   m2   M3

half steps: 2   2   2   1   2   3
intervals: M2   M2   M2   m2   M2   m3

half steps: 2   2   1   2   2   3
intervals: M2   M2   m2   M2   M2   m3

half steps: 2   1   2   2   1   4
intervals: M2   m2   M2   M2   m2   M3

half steps: 1   2   2   1   2   4
intervals: m2   M2   M2   m2   M2   M3

How many distinct interval patterns appear in the group? ___6___

d. E major

half steps: 2   3   2   2   2   1
intervals: M2   m3   M2   M2   M2   m2

half steps: 2   3   2   2   1   2
intervals: M2   m3   M2   M2   m2   M2

half steps: 1   4   2   1   2   2
intervals: m2   M3   M2   m2   M2   M2

half steps: 2   4   1   2   2   1
intervals: M2   M3   m2   M2   M2   m2

half steps: 2   3   2   2   1   2
intervals: M2   m3   M2   M2   m2   M2

half steps: 2   3   2   1   2   2
intervals: M2   m3   M2   m2   M2   M2

half steps: 1   4   1   2   2   2
intervals: m2   M3   m2   M2   M2   M2

How many distinct interval patterns appear in the group? ___6___

EXERCISE 2.5 Write a different two-note series (or melodic interval) for each group, then transpose each series diatonically so that a series begins on each note of the corresponding major scale. Identify all intervals, and for each group determine how many distinct interval patterns appear.

a. C major, seconds

half steps:
intervals:

half steps:
intervals:

half steps:
intervals:

half steps:
intervals:

half steps:
intervals:

half steps:
intervals:

half steps:
intervals:

How many interval patterns? _____

b. C major, thirds

half steps:
intervals:

half steps:
intervals:

half steps:
intervals:

half steps:
intervals:

half steps:
intervals:

half steps:
intervals:

half steps:
intervals:

How many interval patterns? _____

c. C major, fourths

half steps:
intervals:

half steps:
intervals:

half steps:
intervals:

half steps:
intervals:

half steps:
intervals:

half steps:
intervals:

half steps:
intervals:

How many interval patterns? _____

d. C major, fifths

half steps:
intervals:

half steps:
intervals:

half steps:
intervals:

half steps:
intervals:

half steps:
intervals:

half steps:
intervals:

half steps:
intervals:

How many interval patterns? _____

e. C major, sixths

half steps:
intervals:

half steps:
intervals:

half steps:
intervals:

half steps:
intervals:

half steps:
intervals:

half steps:
intervals:

half steps:
intervals:

How many interval patterns? _____

f. C major, sevenths

half steps:
intervals:

half steps:
intervals:

half steps:
intervals:

half steps:
intervals:

half steps:
intervals:

half steps:
intervals:

half steps:
intervals:

How many interval patterns? _____

Interval Patterns and d Distances

Your results for this exercise could have been anticipated from our earlier work in Chapter 1. We determined in Exercise 1.6d (on p. 32) that all interval types (d distances, or distances measured in dots) in the diatonic collection have exactly two qualities (c distances, or distances measured in crossing lines). Therefore, it follows that our group of transposed two-note series (which keep the d distances constant within each group) would have two different interval patterns (two different c distances). By comparing your work in Exercises 2.5 and 1.6d (p. 32), you can deepen your understanding of intervals in the diatonic collection. Solution 2.5 displays the answers to this exercise for your convenience in evaluating your own work.

Inversion and Two-Note Series

After completing Exercise 2.5, you also might have recognized that completing only the first three of these groups of transposed series was sufficient to verify our hypothesis for two-note series, because the last three groups are redundant due to inversion. The interval type of a seventh projected in Exercise 2.5f is the inversion of the interval type of a second in Exercise 2.5a. Likewise, the interval type of a sixth in Exercise 2.5e is the inversion of the interval type of a third in Exercise 2.5b, and the interval type of a fifth in Exercise 2.5d is the inversion of the interval type of a fourth in Exercise 2.5c. Therefore, the interval patterns formed by these pairs of groups are directly related. Because major and minor intervals invert into each other and because inversionally related interval types sum to nine, the interval patterns shown in Solution 2.5a (M2–m7 and m2–M7) invert in Solution 2.5f to the opposite interval patterns (m7–M2 and M7–m2). In terms of half steps, the intervals are also reversed in these pairs so that the numbers of half steps in each pair sum to twelve. Thus, the order of half steps in the interval patterns of Solution 2.5a (2–10 and 1–11) is reversed in Solution 2.5f (10–2 and 11–1). Likewise, the interval patterns in Solution 2.5b and in Solution 2.5c are reversed in their inversional counterparts shown in Solution 2.5e and in Solution 2.5d, respectively. In addition to these inversional pairs, the intervals in each series are also inversionally related because each pair of intervals completes the octave. For example, in Solution 2.5a all intervals of 2 half steps require 10 half steps to complete the octave, and all intervals of 1 half step require 11 half steps to complete the octave. 2–10 and 1–11 are inversionally related, as mentioned previously. The concept of inversional pairs will become an increasingly important component of your studies in music theory, and perhaps observing these pairs in these contexts may help illuminate the concept of interval inversion.

Transposed series of notes drawn from major scales and the resulting interval patterns

a. C major, seconds

half steps:	2	10
intervals:	M2	m7

half steps:	2	10
intervals:	M2	m7

half steps:	1	11
intervals:	m2	M7

half steps:	2	10
intervals:	M2	m7

half steps:	2	10
intervals:	M2	m7

half steps:	2	10
intervals:	M2	m7

half steps:	1	11
intervals:	m2	M7

How many interval patterns? ___2___

b. C major, thirds

half steps:	4	8
intervals:	M3	m6

half steps:	3	9
intervals:	m3	M6

half steps:	3	9
intervals:	m3	M6

half steps:	4	8
intervals:	M3	m6

half steps:	4	8
intervals:	M3	m6

half steps:	3	9
intervals:	m3	M6

half steps:	3	9
intervals:	m3	M6

How many interval patterns? ___2___

c. C major, fourths

half steps :	5	7
intervals:	P4	P5

half steps:	5	7
intervals:	P4	P5

half steps:	5	7
intervals:	P4	P5

half steps:	6	6
intervals:	A4	d5

half steps:	5	7
intervals:	P4	P5

half steps:	5	7
intervals:	P4	P5

half steps:	5	7
intervals:	P4	P5

How many interval patterns? ___2___

d. C major, fifths

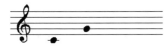

half steps: 7   5
intervals: P5   P4

half steps: 7   5
intervals: P5   P4

half steps: 7   5
intervals: P5   P4

half steps: 7   5
intervals: P5   P4

half steps: 7   5
intervals: P5   P4

half steps: 7   5
intervals: P5   P4

half steps: 6   6
intervals: d5   A4

How many interval patterns? __2__

e. C major, sixths

half steps: 9   3
intervals: M6   m3

half steps: 9   3
intervals: M6   m3

half steps: 8   4
intervals: m6   M3

half steps: 9.   3
intervals: M6   m3

half steps: 9   3
intervals: M6   m3

half steps: 8   4
intervals: m6   M3

half steps: 8   4
intervals: m6   M3

How many interval patterns? __2__

f. C major, sevenths

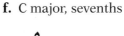

half steps: 11   1
intervals: M7   m2

half steps: 10   2
intervals: m7   M2

half steps: 10   2
intervals: m7   M2

half steps: 11   1
intervals: M7   m2

half steps: 10   2
intervals: m7   M2

half steps: 10   2
intervals: m7   M2

half steps: 10   2
intervals: m7   M2

How many interval patterns? __2__

Cardinality Equals Variety

It seems safe to conclude that the number of notes in a series indicates the number of distinct interval patterns formed by diatonic transpositions of that series, as we originally hypothesized. The formal term for this phenomenon is *cardinality equals variety*. *Cardinality* generally refers to the number of items contained within a set; here, the term refers to the number of notes in a series. *Variety,* in this case, refers to the number of different interval patterns that can be formed by transpositions of the initial series, or the variety of interval patterns. The fact that cardinality equals variety for any series of any number of notes drawn from a diatonic collection is another important property of the diatonic collection—along with the fact that the diatonic collection is maximally even and that it is a deep scale, as we determined in Chapter 1.

So far in this chapter, we have largely taken for granted one important aspect of our observations: we only have been working with diatonic collections. To complete our study of the property cardinality equals variety, we need to examine interval patterns formed by series of notes contained in *other* collections, to determine if this property occurs regardless of the collection of notes from which these series are drawn. In the next section we will test our hypothesis using other scales.

INTERVAL PATTERNS IN OTHER SCALES

Variety in Harmonic and Ascending Melodic Minor

We begin our study of interval patterns in other scales using two of the most familiar scales, at least to students of tonal music, which are not associated with the diatonic collection (major, natural minor, and the modes). We have already established that the harmonic and ascending melodic minor scales are not maximally even; however, the lack of this property in these scales may or may not be relevant to what interval patterns are produced by transpositions of various series of notes drawn from these scales. Exercise 2.6 provides an initial opportunity to test the harmonic and ascending melodic minor scales, respectively, to see if cardinality equals variety for series of notes drawn from these scales. We will begin with a simple series of four consecutive notes from each scale; if cardinality equals variety for these series, we will continue with series of other lengths and complexities in another exercise. In addition to our main purpose in completing this exercise, working with these important scales also will provide us with an excellent opportunity to observe their intervallic structures. Therefore, in Exercise 2.6 determine the interval patterns (and complete the octave) for all of the transpositions of each series—even if not *all* patterns are needed to satisfy your test for the property of cardinality equals variety.

EXERCISE 2.6 For each group, transpose the series diatonically so that a series begins on each note of the corresponding scale. Identify the intervals between each pair of adjacent notes as indicated. For each group, determine how many distinct interval patterns appear.

a. D harmonic minor

half steps:
intervals:

half steps:
intervals:

half steps:
intervals:

half steps:
intervals:

half steps:
intervals:

half steps:
intervals:

half steps:
intervals:

How many distinct interval patterns appear in the group? _____

b. E ascending melodic minor

half steps:
intervals:

half steps:
intervals:

half steps:
intervals:

half steps:
intervals:

half steps:
intervals:

half steps:
intervals:

half steps:
intervals:

How many distinct interval patterns appear in the group? _____

Solution 2.6 shows the completed transposed series and interval patterns; as usual, carefully check your transposed series and interval identifications. As you may have concluded, because more than four interval patterns are formed by transpositions of a four-note series in each scale, cardinality does not equal variety for these scales. There is no need to examine other series of notes drawn from these scales in this regard, because if the property cardinality equals variety applies, it must hold true for all series of any length or configuration drawn from a scale.

Harmonic Minor

In the harmonic minor scale (as shown in Solution 2.6a), no fewer than seven different interval patterns are formed, one for every transposed series. Thus, every group of four consecutive notes drawn from this scale will sound different. In contrast, because cardinality equals variety, the diatonic collection exhibits only four different interval patterns, and some patterns occur more than once (as demonstrated in Exercise 2.2a earlier in this chapter). Perhaps most striking in the interval patterns of the harmonic minor scale is the presence of the augmented second in no fewer than three of these series—a thorny interval, as we discussed in Chapter 1.

Ascending Melodic Minor

The ascending melodic minor scale (as shown in Solution 2.6b), on the other hand, contains five interval patterns among its transposed series of four consecutive notes. Perhaps most notable here is the fact that two interval patterns each contain three consecutive major seconds. These series come close to approximating the whole-tone scale, with its inherent ambiguity, which we explored at the piano in Chapter 1. Taken together, these two series drawn from the ascending melodic minor would overlap to form four consecutive major seconds. Only the presence of minor seconds surrounding this long series of whole steps prevents this scale from approaching the total ambiguity offered by the whole-tone scale. Yet, the fact that this scale contains such a large number of consecutive whole steps suggests that composers must have had to treat this scale, as well as the harmonic minor scale with its problematic augmented second, with care in executing scalar passages.

Transposed series of notes drawn from forms of the minor scale and the resulting interval patterns

a. D harmonic minor

half steps:	2	1	2	7
intervals:	M2	m2	M2	P5

half steps:	1	2	2	7
intervals:	m2	M2	M2	P5

half steps:	2	2	1	7
intervals:	M2	M2	m2	P5

half steps:	2	1	3	6
intervals:	M2	m2	A2	d5

half steps:	1	3	1	7
intervals:	m2	A2	m2	P5

half steps:	3	1	2	6
intervals:	A2	m2	M2	d5

half steps:	1	2	1	8
intervals:	m2	M2	m2	A5

How many distinct interval patterns appear in the group? ____7____

b. E ascending melodic minor

half steps:	2	1	2	7
intervals:	M2	m2	M2	P5

half steps:	1	2	2	7
intervals:	m2	M2	M2	P5

half steps:	2	2	2	6
intervals:	M2	M2	M2	d5

half steps:	2	2	2	6
intervals:	M2	M2	M2	d5

half steps:	2	2	1	7
intervals:	M2	M2	m2	P5

half steps:	2	1	2	7
intervals:	M2	m2	M2	P5

half steps:	1	2	1	8
intervals:	m2	M2	m2	A5

How many distinct interval patterns appear in the group? ____5____

Minor Contexts

Because the harmonic minor and ascending melodic minor scales are not maximally even, and because cardinality does not equal variety for series of notes drawn from these scales, you may be forming the opinion, by now, that these scales are somewhat inferior to those scales that correspond to the diatonic collection where these properties are in force. In some sense your conclusion would be right. These scales are indeed subordinate to the diatonic collection because they draw their tonal center from, and are usually defined in reference to, the natural minor scale. "Raise the seventh note, and raise the sixth and seventh notes" are the traditional directions for constructing these two scales from the natural minor scale. Nevertheless, composers seem to have been extremely attracted to these scales; the harmonic and melodic minor scale forms became the norms for tonal music, not the natural minor scale from which they derive.

In light of this apparent disparity between theory and practice, how are we to view our conclusions regarding maximal evenness and cardinality equals variety? Does the frequency with which these forms of the minor scale are employed by tonal composers negate the importance of our theoretical excursions? Quite the contrary. We have already shown the importance of maximally even structures to our perception of the conspicuous gap between steps in the harmonic minor scale. Moreover, what our findings about these minor scale forms have shown, and what we must keep in mind regarding these scales, is that these scales represent a compositional compromise. Tonal composers seem to have viewed the internal structure of collections, or scales, as subordinate to matters of voice leading (the behavior of individual voices or lines within a musical passage). Therefore, they chose to raise the seventh note of the scale, in harmonic minor, so that this seventh note of the scale could lead by a half step up to the tonic goal of the scale. Raising the sixth note of the scale, in melodic minor, simply smoothes out the perceived gap in the harmonic minor scale, the augmented second mentioned earlier. Composers seem to have compromised the maximal evenness of the natural minor scale, in addition to cardinality equals variety, to create a stronger voice-leading approach to the tonic. As you continue your study of tonal music, problems of voice leading, especially in minor keys, will become an increasingly important concern. The theoretical foundations laid here are intended to enhance your appreciation for this facet of tonal music.

Of course, composers of tonal music in the common-practice period literally knew nothing about maximal evenness or cardinality equals variety, though they may have had intuitive notions along these lines that they did not express; therefore, my assertions regarding compromises made by these composers must be viewed in this limited context. We cannot know the *reasons* for choices made by these composers, but our theoretical inquiries can shape or reshape *our perception* of the results of their compositional choices. Therefore, developing an understanding of the structure of the diatonic collection by means of these theoretical undertakings provides us with a deeper appreciation of the subtle differences among the variants of the minor scale (natural, harmonic, and melodic).

Maximal Evenness and Cardinality Equals Variety

So far we have ascertained that cardinality equals variety for series of notes drawn from the diatonic collection, which *is* maximally even, and that cardinality does not equal variety for series of notes drawn from the harmonic and ascending melodic minor scales, which *are not* maximally even. We continue our study of the property cardinality equals variety by examining interval patterns formed by series of notes drawn from other maximally even scales. In this way we can determine if the two properties are intertwined.

Pentatonic, Whole Tone, and Octatonic

Exercise 2.7 contains groups of simple series of three consecutive notes drawn from the pentatonic, whole tone, and octatonic scales—the other three maximally even scales that we studied in Chapter 1. All of the series are provided for you, because transposing series within these scales is not as straightforward on the staff as within seven-note scales. Also, only half steps are calculated, because these scales do not conform to the configuration of the familiar staff—therefore, naming intervals by quality and type may produce misleading results. In terms of circle diagrams, we still keep the d distances constant as we rotate (or transpose) the patterns around the circle, while we observe the c distances between the selected dots in the resulting patterns. However, we record only the c distances (half steps), rather than include the interval types and qualities, because the d distances do not correspond directly to traditional interval types (seconds, thirds, and so forth) as in seven-note collections. Again, complete the octave for each transposed series either by showing or assuming a parenthetical note an octave above each initial note. By examining the interval patterns formed by these three-note series drawn from each of these scales, we will obtain a preliminary impression of whether or not cardinality equals variety for these scales. If any of these scales do exhibit the property cardinality equals variety, we will test other series of notes to examine the property fully.

For each group, identify the intervals between each pair of adjacent notes as indicated, and determine how many distinct interval patterns appear.

a. Pentatonic

b. Whole tone

half steps:

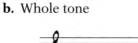

half steps:

half steps:

half steps:

half steps:

half steps:

half steps:

half steps:

half steps:

half steps:

half steps:

How many distinct interval patterns appear in the group? _____

How many distinct interval patterns appear in the group? _____

c. Octatonic

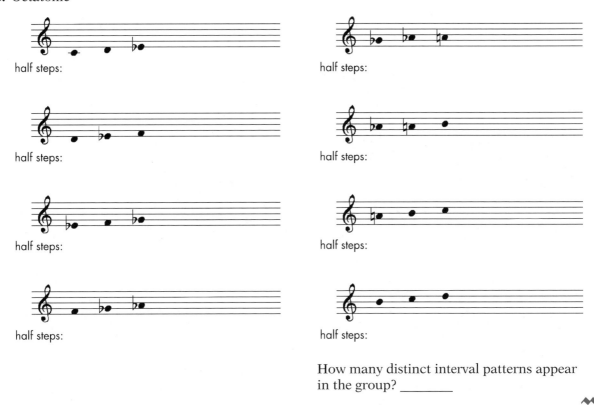

half steps:

half steps:

half steps:

half steps:

half steps:

half steps:

half steps:

half steps:

How many distinct interval patterns appear in the group? _____

As shown in Solution 2.7, only the pentatonic scale exhibits cardinality equals variety for the series of notes we examined. The other two scales each have *fewer* than three interval patterns for their groups of three-note series, unlike the harmonic and ascending melodic minor scales which have greater than three distinct patterns.

The Whole-Tone Scale

The whole-tone scale (as shown in Solution 2.7b) remarkably has only a single interval pattern for its group of three-note series. If you wish, you can easily check series of any other lengths or configurations and ascertain that *any* group of series drawn from the whole-tone scale will have only a *single* interval pattern, due to the special configuration of this scale. Therefore, any series of notes drawn from the whole-tone scale will sound almost exactly the same as any other similarly configured series of notes. This ambiguous aspect of the whole-tone scale is perhaps one of the principal charms that attracted composers such as Claude Debussy, who apparently sought a less goal-directed musical palette.

The Octatonic Scale

The octatonic scale (as shown in Solution 2.7c)—also a remarkably consistent scale in terms of spacing, as we observed in Chapter 1—produces only two interval patterns for the group of series consisting of three consecutive notes. This scale, similarly, has *exactly two* interval patterns, regardless of the structure or length of the series. As in the whole-tone scale, the consistent configuration of the octatonic scale makes it difficult to determine where these patterns appear in relation to the tonic note of the scale. Although there is less ambiguity than in the whole-tone scale, such features seem to have begun to become attractive to composers in the early twentieth century.

SOLUTION 2.7 Transposed series of notes drawn from various scales and the resulting interval patterns

a. Pentatonic

half steps: 2   2   8

half steps: 2   3   7

half steps: 3   2   7

half steps: 2   3   7

half steps: 3   2   7

How many distinct interval patterns appear in the group? ___3___

b. Whole tone

half steps: 2   2   8

half steps: 2   2   8

half steps: 2   2   8

half steps: 2   2   8

half steps: 2   2   8

half steps: 2   2   8

How many distinct interval patterns appear in the group? ___1___

c. Octatonic

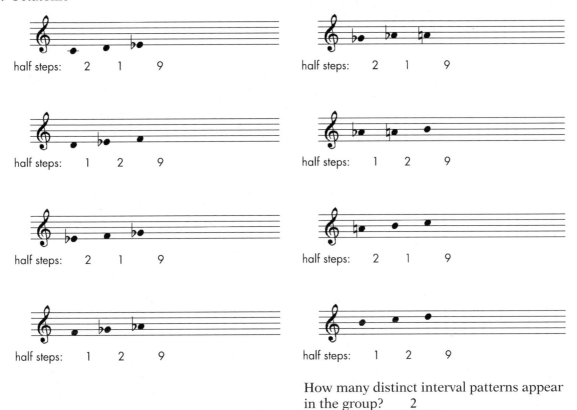

half steps: 2 1 9

half steps: 2 1 9

half steps: 1 2 9

half steps: 1 2 9

half steps: 2 1 9

half steps: 2 1 9

half steps: 1 2 9

half steps: 1 2 9

How many distinct interval patterns appear in the group? _____2_____

The Pentatonic Scale

On the other hand, the pentatonic scale (as shown in Solution 2.7a), like the diatonic collection examined earlier in the chapter, has three different interval patterns corresponding to the group of three-note series—an exact correspondence between the number of notes and the number of patterns. However, to determine if cardinality equals variety for this scale, we must examine some additional groups of series to be sure that the property continues to hold. Exercise 2.8 provides a structure for this investigation. Because no potentially misleading results will be obtained by using the more traditional interval identification system with this scale, we will return to the use of both half steps and interval types and qualities for this exercise. In this way you can obtain some additional interval identification practice using the traditional system of interval types and qualities. Although these examples will not be sufficient to *prove* cardinality equals variety for the pentatonic scale, perhaps these examples will be sufficient to be convincing.[2]

EXERCISE 2.8   For each group drawn from the pentatonic scale, identify the intervals between each pair of adjacent notes as indicated, and determine how many distinct interval patterns appear.

a. Pentatonic

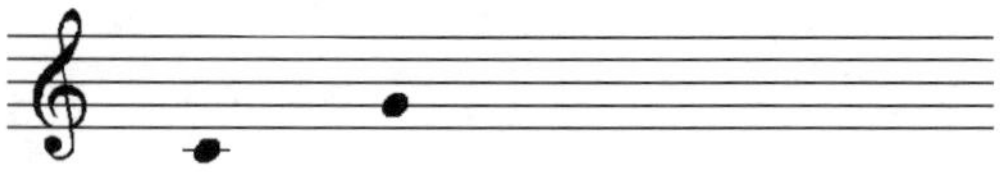

half steps:
intervals:

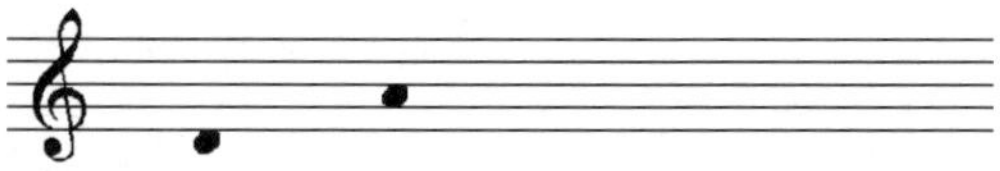

half steps:
intervals:

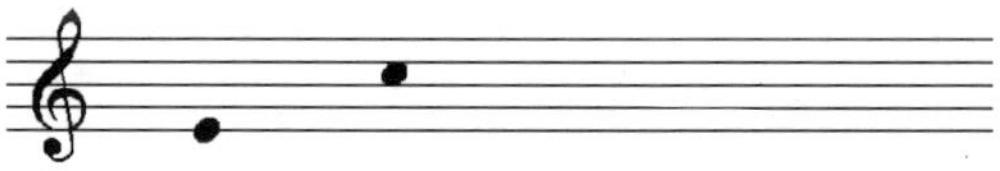

half steps:
intervals:

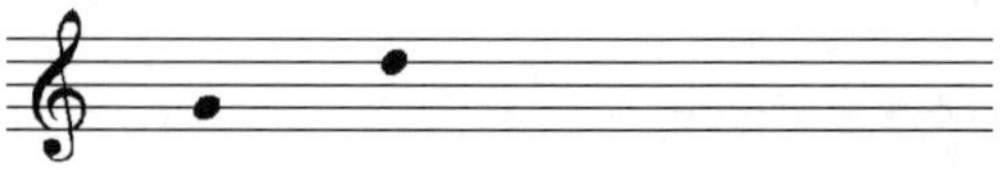

half steps:
intervals:

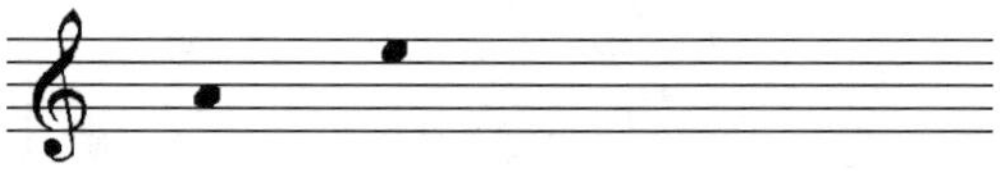

half steps:
intervals:

How many distinct interval patterns appear in the group? _____

b. Pentatonic

half steps:
intervals:

half steps:
intervals:

half steps:
intervals:

half steps:
intervals:

half steps:
intervals:

How many distinct interval patterns appear in the group? _____

c. Pentatonic

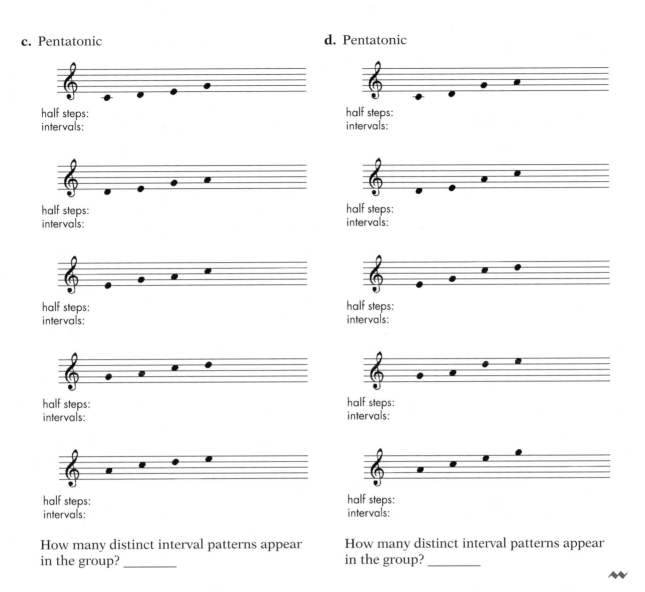

half steps:
intervals:

half steps:
intervals:

half steps:
intervals:

half steps:
intervals:

half steps:
intervals:

How many distinct interval patterns appear in the group? _____

d. Pentatonic

half steps:
intervals:

half steps:
intervals:

half steps:
intervals:

half steps:
intervals:

half steps:
intervals:

How many distinct interval patterns appear in the group? _____

For every group of series in the exercise, the number of interval patterns formed exactly matches the number of notes in the series, as shown in Solution 2.8. Cardinality equals variety for the pentatonic scale—though this outcome is hardly surprising. We have already demonstrated in Chapter 1 that the pentatonic scale is the complement of the diatonic scale. Because the pentatonic scale fits into the spaces around the circle between the dots of the diatonic collection, it is reasonable to expect that the two collections would be closely related in other ways as well.

Transposed series of notes drawn from the pentatonic scale and the resulting interval patterns.

a. Pentatonic

half steps:	7	5
intervals:	P5	P4

half steps:	7	5
intervals:	P5	P4

half steps:	8	4
intervals:	m6	M3

half steps:	7	5
intervals:	P5	P4

half steps:	7	5
intervals:	P5	P4

How many distinct interval patterns appear in the group? ____2____

b. Pentatonic

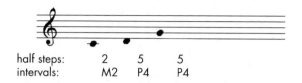

half steps:	2	5	5
intervals:	M2	P4	P4

half steps:	2	5	5
intervals:	M2	P4	P4

half steps:	3	5	4
intervals:	m3	P4	M3

half steps:	2	5	5
intervals:	M2	P4	P4

half steps:	3	4	5
intervals:	m3	M3	P4

How many distinct interval patterns appear in the group? ____3____

c. Pentatonic

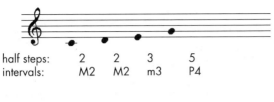

half steps: 2   2   3   5
intervals: M2   M2   m3   P4

half steps: 2   3   2   5
intervals: M2   m3   M2   P4

half steps: 3   2   3   4
intervals: m3   M2   m3   M3

half steps: 2   3   2   5
intervals: M2   m3   M2   P4

half steps: 3   2   2   5
intervals: m3   M2   M2   P4

How many distinct interval patterns appear in the group? ___4___

d. Pentatonic

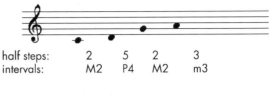

half steps: 2   5   2   3
intervals: M2   P4   M2   m3

half steps: 2   5   3   2
intervals: M2   P4   m3   M2

half steps: 3   5   2   2
intervals: m3   P4   M2   M2

half steps: 2   5   2   3
intervals: M2   P4   M2   m3

half steps: 3   4   3   2
intervals: m3   M3   m3   M2

How many distinct interval patterns appear in the group? ___4___

INTERVAL PATTERNS AND THE CIRCLE OF FIFTHS

After completing the exercises so far in this chapter and comparing the various interval patterns formed in series drawn from the diatonic collection, you have seen that not all interval patterns occur the same number of times in a group. Some interval patterns occur multiple times, whereas others appear only once. We shall now focus on this observation, and examine these frequencies of occurrence in more detail to attempt to determine if there is any recognizable design in the number of occurrences of each interval pattern within a group.

Seven-Line Circle Diagrams

In order to carry out this examination, we will use circle diagrams that are significantly different from the ones we worked with in Chapter 1. These circle diagrams have only seven lines crossing the circle, representing the seven notes of the diatonic collection. Furthermore, the lines are labeled according to the circle-of-fifths model, rather than ascending stepwise through the scale. Thus, the note names assigned to the crossing lines increase by the interval of a fifth as we move clockwise around the circle, as shown in Figure 2.1d.[3]

Figure 2.1 Demonstration of a generated collection. The first three circles (a–c) show the gradual formation of the diatonic collection by means of a constant generating c distance (7). The last circle (d) shows the generating c distance (7) between adjacent lines, rather than a circle of half steps as in previous diagrams.

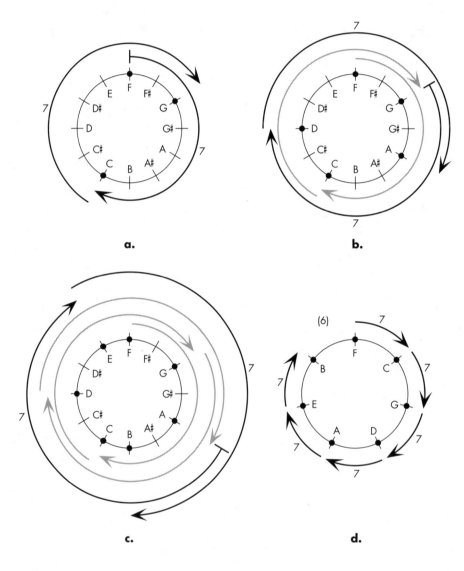

The Generator

The two different kinds of circle diagrams may be reconciled through an elementary understanding of a *generated* collection.[4] A generated collection is formed by repeatedly adding a constant c distance (a single number) around a chromatic (twelve-line) circle until an entire collection is formed. All deep scales (such as the diatonic collection, as mentioned in Chapter 1) can be generated by any interval (g) that is coprime with c (the number of lines around the circle).[5] Integers are termed *coprime* if they have a greatest common divisor of one—or in formulaic terms, the GCD of (g, c) = 1; in this case GCD (7, 12) = 1 and GCD (5, 12) = 1. Thus, as shown in Figure 2.1, the diatonic collection can be generated by a c distance of 7 (or also by 5). The first circle diagram (Figure 2.1a), beginning with the note F and employing the original chromatically oriented twelve-line circle, shows two moves via the generator, depicted by the arrows. The second and third circle diagrams (Figures 2.1b and 2.1c) each show two more steps through the process. The final circle diagram (Figure 2.1d) shows the new seven-line circle that results from this generative process. This circle begins with F, and the lines around the circle are labeled successively by intervals of 7 (g = 7). Thus, the lines are labeled F–C–G–D–A–E–B, with each new note 7 semitones away from the previous one, as shown in Figure 2.1d and as generated in Figures 2.1a, b, and c.

Furthermore, any generated collection can be generated by g or by c − g (the total number of lines around the circle minus the constant generator). Consequently, and as the GCD formula likewise reveals, the diatonic collection also can be generated by a c distance of 5, the inversion of a c distance of 7 (12 − 7 = 5, or 12 − 5 = 7).

Well Formed

Moreover, a generated collection where a single d distance corresponds to the c distance of the generator is called *well formed*.[6] In Figure 2.1c the constant c distance used as a generator (7) corresponds to a single d distance of 4, or in musical terms, a fifth. Thus, the diatonic collection is well formed and is generated by a circle of perfect fifths—for example, F to C, C to G, G to D, D to A, A to E, and E to B are all perfect fifths (c distance of 7, d distance of 4).[7]

The converse, on the other hand, is not always true: In the diatonic collection all fifths (d distances of 4) are not perfect (c distances of 7). However, for collections that are maximally even and well formed, there never will be more than *one* c distance corresponding to the d distance of the generator that does not match the generator (g). And this single c distance will be the distance required to complete the circle and will always be g ± 1 (the generating c distance plus or minus one), if c and d are coprime. In the maximally even and well-formed diatonic collection, c and d are coprime, GCD of (12, 7) = 1. Thus, only one d distance of 4, the one that occurs between B and F, has a c distance of 6 (g ± 1, or 7 − 1). This single interval, the fifth required to complete the circle, corresponds to the diminished fifth (or tritone), as shown between adjacent lines in Figure 2.1d. All of the other fifths are perfect (c distance of 7).[8]

The Generated and Well-Formed Diatonic Collection

To summarize, the diatonic collection may be *generated* by a c distance of 7, these c distances are all fifths (making the diatonic collection *well formed*), and the *single* interval needed to complete the circle is a diminished fifth (a c distance of 6, g ± 1, or 7 − 1). The following exercises display the diatonic collection as generated in this way by the circle of fifths on a seven-line circle diagram. In these exercises the tonic note of the scale under investigation, rather than the initial note of the generation procedure, is located at the top of the circle.

A Table of Observations

Exercise 2.9 provides a framework for the comparison of interval patterns formed by the various groups of series that were drawn from the diatonic collection in exercises presented earlier in this chapter. To complete this new exercise, return to Exercises 2.1, 2.2, and 2.4, and for each of the groups in these earlier exercises, list the initial series used, the diatonic collection from which the series is drawn (labeled according to the major scale that corresponds with the key signature), the distinct interval patterns formed (shown as patterns of half steps), and the number of times that each of these interval patterns occurs in the group. Record these observations in Exercise 2.9, as demonstrated in the sample provided. For example, the group of transposed series in Exercise 2.1a has an initial series of C–D–E, drawn from the diatonic collection corresponding to the C major scale, and has three interval patterns (2–2–8, which occurs three times; 2–1–9, which occurs two times; and 1–2–9, which also occurs twice).

Plotting Series and Observing Distances

Finally, plot the initial series of each group on the seven-line circle diagram provided and determine the distances between adjacent dots in terms of the number of crossing lines around the circle of fifths. Place these distances into the appropriate box in the table in any order; this box is located directly above the circle diagram. For example, using the group of series from Exercise 2.1a, the first circle diagram in Exercise 2.9 is labeled beginning at the top of the circle with C and continues clockwise around the circle with G, D, A, E, B, and F, ascending through the diatonic circle of fifths. The distances between adjacent dots around the circle described by this pattern are 2, 2, and 3, as tabulated in the box directly above the circle. The alterations to the circle diagrams described earlier and the representation of the corresponding series of notes in reference to the circle of fifths will allow us to observe additional aspects of the series that we were unable to observe in the circles with twelve crossing lines in reference to the chromatic scale. Complete Exercise 2.9 based on these instructions and on the sample solution provided.

Using Exercises 2.1, 2.2, and 2.4 from earlier in the chapter, plot the initial series on the circle diagrams and complete the tables as indicated.

Group of series from exercise	Initial series of notes	Drawn from which collection	Distances between dots around circle of fifths
2.1a (p. 47)	C–D–E	C major	2, 2, 3

Interval pattern	Number of occurrences
2–2–8	3
2–1–9	2
1–2–9	2

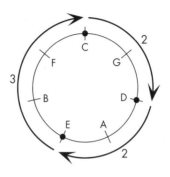

Group of series from exercise	Initial series of notes	Drawn from which collection	Distances between dots around circle of fifths
2.1b			

Interval pattern	Number of occurrences

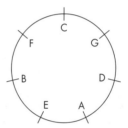

Group of series from exercise	Initial series of notes	Drawn from which collection	Distances between dots around circle of fifths
2.1c			

Interval pattern	Number of occurrences

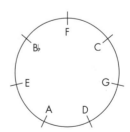

Group of series from exercise	Initial series of notes	Drawn from which collection	Distances between dots around circle of fifths
2.1d			

Interval pattern	Number of occurrences		

Group of series from exercise	Initial series of notes	Drawn from which collection	Distances between dots around circle of fifths
2.2a (p. 50)			

Interval pattern	Number of occurrences		

Group of series from exercise	Initial series of notes	Drawn from which collection	Distances between dots around circle of fifths
2.2b			

Interval pattern	Number of occurrences		

Group of series from exercise	Initial series of notes	Drawn from which collection	Distances between dots around circle of fifths
2.2c			

Interval pattern	Number of occurrences

Group of series from exercise	Initial series of notes	Drawn from which collection	Distances between dots around circle of fifths
2.2d			

Interval pattern	Number of occurrences

Group of series from exercise	Initial series of notes	Drawn from which collection	Distances between dots around circle of fifths
2.4a (p. 57)			

Interval pattern	Number of occurrences

Group of series from exercise	Initial series of notes	Drawn from which collection	Distances between dots around circle of fifths
2.4b			

Interval pattern	Number of occurrences	

Group of series from exercise	Initial series of notes	Drawn from which collection	Distances between dots around circle of fifths
2.4c			

Interval pattern	Number of occurrences	

Group of series from exercise	Initial series of notes	Drawn from which collection	Distances between dots around circle of fifths
2.4d			

Interval pattern	Number of occurrences	

Forming a Hypothesis

Solution 2.9 shows the completed tables and circle diagrams. Carefully check your answers against those provided and make any needed corrections. Then, continue to examine Exercise 2.9, comparing the number of occurrences of each interval pattern with the distances between notes around the circle of fifths. Based on your observations in making this comparison, create a generalized statement that relates the distances with the number of occurrences of each pattern. Record your hypothesis in whatever words seem appropriate, but as accurately and specifically as you can, in Exercise 2.10.

SOLUTION 2.9 The initial series plotted on the circle diagrams and the completed tables for each indicated exercise from earlier in the chapter

Group of series from exercise	Initial series of notes	Drawn from which collection	Distances between dots around circle of fifths
2.1a	C–D–E	C major	2, 2, 3

Interval pattern	Number of occurrences
2–2–8	3
2–1–9	2
1–2–9	2

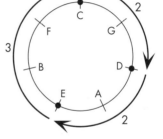

Group of series from exercise	Initial series of notes	Drawn from which collection	Distances between dots around circle of fifths
2.1b	C–D–F	C major	2, 4, 1

Interval pattern	Number of occurrences
2–3–7	4
1–4–7	2
2–4–6	1

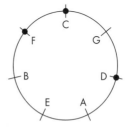

Group of series from exercise	Initial series of notes	Drawn from which collection	Distances between dots around circle of fifths
2.1c	F–G–C	F major	1, 1, 5

Interval pattern	Number of occurrences
2–5–5	5
1–6–5	1
1–5–6	1

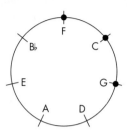

Group of series from exercise	Initial series of notes	Drawn from which collection	Distances between dots around circle of fifths
2.1d	E♭–G–A♭	E♭ major	4, 2, 1

Interval pattern	Number of occurrences
4–1–7	2
3–2–7	4
4–2–6	1

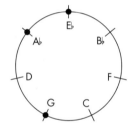

Group of series from exercise	Initial series of notes	Drawn from which collection	Distances between dots around circle of fifths
2.2a	C–D–E–F	C major	2, 2, 2, 1

Interval pattern	Number of occurrences
2–2–1–7	2
2–1–2–7	2
1–2–2–7	2
2–2–2–6	1

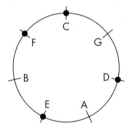

Group of series from exercise	Initial series of notes	Drawn from which collection	Distances between dots around circle of fifths
2.2b	D–E–F♯–A	D major	1, 1, 2, 3

Interval pattern	Number of occurrences	
2–2–3–5	3	
2–1–4–5	2	
1–2–4–5	1	
1–2–3–6	1	

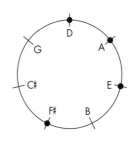

Group of series from exercise	Initial series of notes	Drawn from which collection	Distances between dots around circle of fifths
2.2c	F–G–B♭–C	F major	1, 1, 4, 1

Interval pattern	Number of occurrences	
2–3–2–5	4	
1–4–2–5	1	
2–4–1–5	1	
1–4–1–6	1	

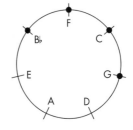

Group of series from exercise	Initial series of notes	Drawn from which collection	Distances between dots around circle of fifths
2.2d	B♭–D–E♭–G	B♭ major	3, 1, 2, 1

Interval pattern	Number of occurrences	
4–1–4–3	2	
3–2–4–3	1	
3–2–3–4	3	
4–2–3–3	1	

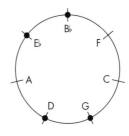

Group of series from exercise	Initial series of notes	Drawn from which collection	Distances between dots around circle of fifths
2.4a	C–D–E–F–G	C major	1, 1, 2, 2, 1

Interval pattern	Number of occurrences	
2–2–1–2–5	2	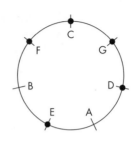
2–1–2–2–5	2	
1–2–2–2–5	1	
2–2–2–1–5	1	
1–2–2–1–6	1	

Group of series from exercise	Initial series of notes	Drawn from which collection	Distances between dots around circle of fifths
2.4b	D♭–E♭–F–A♭–C	D♭ major	1, 1, 2, 1, 2

Interval pattern	Number of occurrences	
2–2–3–4–1	2	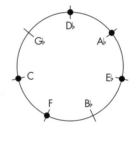
2–1–4–3–2	2	
1–2–4–3–2	1	
2–2–3–3–2	1	
1–2–3–4–2	1	

Group of series from exercise	Initial series of notes	Drawn from which collection	Distances between dots around circle of fifths
2.4c	C–D–E–F–G–A	C major	1, 1, 1, 1, 2, 1

Interval pattern	Number of occurrences	
2–2–1–2–2–3	2	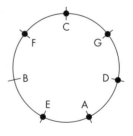
2–1–2–2–2–3	1	
1–2–2–2–1–4	1	
2–2–2–1–2–3	1	
2–1–2–2–1–4	1	
1–2–2–1–2–4	1	

Group of series from exercise	Initial series of notes	Drawn from which collection	Distances between dots around circle of fifths
2.4d	E–F♯–A–B–C♯–D♯	E major	1, 1, 1, 2, 1, 1

Interval pattern	Number of occurrences
2–3–2–2–2–1	1
2–3–2–2–1–2	2
1–4–2–1–2–2	1
2–4–1–2–2–1	1
2–3–2–1–2–2	1
1–4–1–2–2–2	1

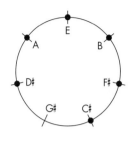

EXERCISE 2.10

Based on your examination of the distances between notes around the circle of fifths and the number of occurrences of the interval patterns formed for each group in Exercise 2.9, make a generalized statement to explain your observation.

_____

_____

_____

_____

Clearly, the distances between adjacent dots around the circle of fifths and the number of occurrences of the interval patterns are intimately related. The group of series from Exercise 2.1b, shown in Solution 2.9, has the same number of occurrences of interval patterns as the distances between the adjacent dots as plotted around the circle of fifths. Likewise, the group of series from Exercise 2.1d has exactly the same number of occurrences and distances between adjacent dots as the group from Exercise 2.1b. Because both of these initial series of notes consist of intervals of a second and a third between adjacent notes, it is easy to see why these different series would be so closely related. The group of series from Exercise 2.1c has different numbers of occurrences of its interval patterns than the previously mentioned pair; nevertheless, these interval patterns again match the distances between adjacent dots around the circle of fifths described by the initial series of notes. These three notes occur as adjacencies in the circle of fifths, therefore two interval patterns occur only once, as suggested by the distances between these adjacent notes on the circle of fifths, and one interval pattern occurs five times.

The larger series of notes tabulated in Solution 2.9 exhibit similar relationships. The series with four notes, originally explored in Exercise 2.2, each have the same numbers of occurrences of interval patterns as distances between adjacent dots around the circle of fifths. Furthermore, the groups of series from Exercises 2.2b and 2.2d have exactly the same numbers of occurrences, again due to their related interval structures, as each contains two seconds and a third between adjacent notes in their initial series. Also, the initial series from Exercise 2.2c again outlines adjacent notes on the circle of fifths, resulting in similar numbers of occurrences of its interval patterns to those we saw in the series from Exercise 2.1c. The other series tabulated in Solution 2.9, with five or six notes, exhibit similar characteristics.

In light of these observations, the generalized statement you have proposed may be somewhat similar to the one provided in Solution 2.10. However, regardless of the words you have chosen to express your observations, your statement should somehow acknowledge that the distances between notes around the circle of fifths indicates the number of times each interval pattern is formed. This assertion is an extremely powerful statement about the diatonic collection. The fact that these correspondences exist for all of these series, as you can easily verify in Solution 2.9, is remarkable.

SOLUTION 2.10

A generalized statement that relates the distances between notes around the circle of fifths with the number of occurrences of the interval patterns formed for each group in Exercise 2.9

The distances between notes around the circle of fifths indicate

the number of times that each interval pattern is formed by all

of the series in a group.

~

Structure Implies Multiplicity

The formal term for this concept is *structure implies multiplicity*. *Structure*, in this particular case, refers to the intervals between notes measured in fifths. *Multiplicity* refers to the number of times that interval patterns appear in the group of transposed series. Similar to the fact that cardinality equals variety, structure implies multiplicity for any series of any number of notes drawn from a diatonic collection.

Next, we will attempt to determine if the other musical structures we studied in connection with cardinality equals variety behave similarly with regard to structure implies multiplicity. If they do, we would expect to find that structure implies multiplicity in the pentatonic scale, whereas in the harmonic minor, ascending melodic minor, octatonic, and whole-tone scales, structure does not imply multiplicity.

The Generated Pentatonic Collection

As shown in Figure 2.2, the pentatonic collection may be formed by the same generators as the diatonic collection—7 and 5, g and c − g, or in musical terms the perfect fifth and its inversion, the perfect fourth. The first two circles (Figures 2.2a and 2.2b) demonstrate the generation of the pentatonic collection, beginning with F and moving around the twelve-line circle in perfect fifths (c = 7) until the five-note collection is complete. The generating c distance again corresponds to a single d distance; thus, the pentatonic

Figure 2.2 Demonstration of a generated collection. The first two circles (a and b) show the gradual formation of the pentatonic collection by means of a constant generating c distance (7). The last circle (c) represents the pentatonic collection as a circle of fifths and shows the generating c distance (7) between adjacent lines.

collection, like the diatonic collection, is well formed. However, because there are fewer dots than in the diatonic case, the generating c distance of 7 corresponds to a d distance of 3. As shown in Figure 2.2c between adjacent lines, and because c and d are coprime for the maximally even and well-formed pentatonic collection, here again *one* d distance of 3 does not match the c distance of the generator: the distance between A and F required to complete the circle is 8 (g ± 1 or 7 + 1).[9]

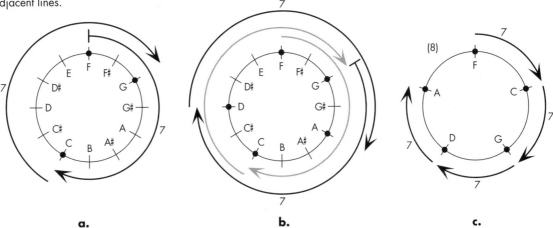

a.   b.   c.

Interval Structures and Interval Patterns

In Exercise 2.11 we will explore the interval structures of and the numbers of interval patterns formed by series of notes drawn from the pentatonic scale that were given initially in Exercises 2.7a and 2.8. As before, the notes are arranged around the circle of perfect fifths, the generating interval. However, the last "fifth" (d distance of 3) from E to C, in *traditional* terms, is a minor sixth. Nevertheless, all of these notes are placed the same distance from one another in the pentatonic sense, because the d distances between each pair of notes are the same. Thus, the circle is closed and complete, despite its appearance in traditional intervallic terms.

EXERCISE 2.11 Using Exercises 2.7a and 2.8 from earlier in the chapter, plot the initial series on the circle diagrams and complete the tables as indicated.

Group of series from exercise	Initial series of notes	Drawn from which collection	Distances between dots around circle of fifths
2.7a (p. 74)			

Interval pattern	Number of occurrences

Group of series from exercise	Initial series of notes	Drawn from which collection	Distances between dots around circle of fifths
2.8a (p. 78)			

Interval pattern	Number of occurrences	

Group of series from exercise	Initial series of notes	Drawn from which collection	Distances between dots around circle of fifths
2.8b			

Interval pattern	Number of occurrences	

Group of series from exercise	Initial series of notes	Drawn from which collection	Distances between dots around circle of fifths
2.8c			

Interval pattern	Number of occurrences	

Group of series from exercise	Initial series of notes	Drawn from which collection	Distances between dots around circle of fifths
2.8d			

Interval pattern	Number of occurrences	

Based on your findings after completing Exercise 2.11, determine if structure implies multiplicity for the pentatonic scale. Then, check your answers to the exercise carefully against those provided in Solution 2.11.

Does structure imply multiplicity for the pentatonic scale?_____

We have already determined that cardinality equals variety for the pentatonic collection and that it is maximally even. Based on our answers in Exercise 2.11, we now can also conclude that structure implies multiplicity for this collection as well.

STRUCTURE IN OTHER COLLECTIONS

We have now seen that the diatonic and pentatonic collections, which exhibit the property cardinality equals variety, also exhibit the property structure implies multiplicity. Next we will examine structures and multiplicities in collections that do not display cardinality equals variety. If the two properties are related, we also would expect to find that structure does not imply multiplicity for these collections.

The Bisector

Unlike the diatonic and pentatonic collections, neither the harmonic minor nor the ascending melodic minor scales can be produced by a generator, or a single c distance applied consecutively around the circle. Therefore, we cannot plot the harmonic minor and ascending melodic minor collections on circles in the same way that we plotted the diatonic and pentatonic collections. However, the same diagrams constructed for the diatonic and pentatonic collections using their generators also could have been plotted using their *bisectors*.[10] The generator is a much more powerful construct (and is intimately related to structure implies multiplicity), thus we will continue to use generators rather than bisectors to build collections whenever possible. However, an understanding of bisectors will help us plot collections that are not generated, such as the ascending melodic minor, harmonic minor, and octatonic.

The Bisector Defined

A bisector divides the octave *approximately* in half. Therefore, in seven-dot circles such as those used for the diatonic collection, the octave is bisected by d distances of 3 or 4 (the two intervals that are approximately half of d, or 7).[11] The diatonic collection generated by constant c distances of 7, as shown in Figure 2.1, also could have been produced by a consistent application of the bisector 4 (successive d distances of 4 dots applied consistently around the circle).[12] This correspondence between generator and bisector occurs because the diatonic collection is well formed, as discussed earlier (the generating c distance corresponds to a single d distance, in this case the bisector). Similarly, the pentatonic collection generated by constant c distances of 7, as shown in Figure 2.2, also could have been produced by the bisector of 3 or 2 (each of which is approximately half of d, or 5). The pentatonic collection also is well formed, for the same reason (a single d distance corresponds to the generator, in this case the bisector as well).

The initial series plotted on the circle diagram and the completed table for each indicated exercise from earlier in the chapter

Group of series from exercise	Initial series of notes	Drawn from which collection	Distances between dots around circle of fifths
2.7a	C–D–E	Pentatonic	2, 2, 1

Interval pattern	Number of occurrences
2–2–8	1
2–3–7	2
3–2–7	2

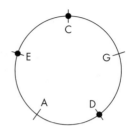

Group of series from exercise	Initial series of notes	Drawn from which collection	Distances between dots around circle of fifths
2.8a	C–G	Pentatonic	1, 4

Interval pattern	Number of occurrences
7–5	4
8–4	1

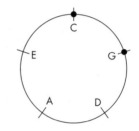

Group of series from exercise	Initial series of notes	Drawn from which collection	Distances between dots around circle of fifths
2.8b	C–D–G	Pentatonic	1, 1, 3

Interval pattern	Number of occurrences
2–5–5	3
3–5–4	1
3–4–5	1

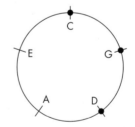

Group of series from exercise	Initial series of notes	Drawn from which collection	Distances between dots around circle of fifths
2.8c	C–D–E–G	Pentatonic	1, 1, 2, 1

Interval pattern	Number of occurrences
2–2–3–5	1
2–3–2–5	2
3–2–3–4	1
3–2–2–5	1

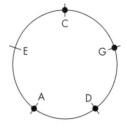

Group of series from exercise	Initial series of notes	Drawn from which collection	Distances between dots around circle of fifths
2.8d	C–D–G–A	Pentatonic	1, 1, 1, 2

Interval pattern	Number of occurrences
2–5–2–3	2
2–5–3–2	1
3–5–2–2	1
3–4–3–2	1

Bisectors and Harmonic and Ascending Melodic Minor

Although the harmonic and ascending melodic minor collections are certainly not well formed, because they are not even generated, these collections can be produced by means of their bisectors (d distances of 3 or 4, or approximately half of 7). The resulting diagrams will somewhat resemble those that we used to examine the diatonic and pentatonic collections, because the d distances in each case correspond to fifths in the traditional sense. Although both the harmonic and ascending melodic minor collections can be produced by a complete, closed circle of fifths, many of the fifths in these scales are not *perfect* fifths: they do not correspond to a single c distance of 7. As shown in Figure 2.3, using a bisector (d distance of 4, as in the diatonic collection) to produce the harmonic and ascending melodic minor collections results in corresponding c distances of 6, 7, and 8 (rather than the constant c distances used to generate the diatonic collection). Nevertheless, we will use circle diagrams constructed by means of bisectors for the harmonic and ascending melodic minor collections in Exercise 2.12, where we will check these diagrams to see if structure implies multiplicity. However, it is important to keep in mind that though the d distances are constant in these resulting circle diagrams, the c distances vary; thus, these collections are not generated and consequently are not well formed.

Figure 2.3 The harmonic and ascending melodic minor collections constructed by their bisectors (4). These collections are not generated, because no constant c distance may be repeatedly added to produce the entire collection; however, the d distances are constant (and correspond to the circle of fifths).

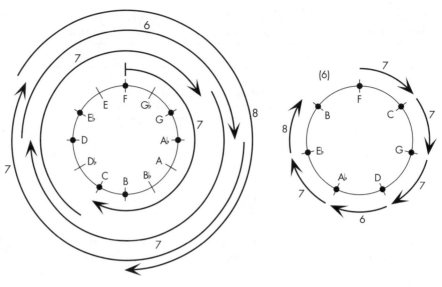

a. harmonic minor

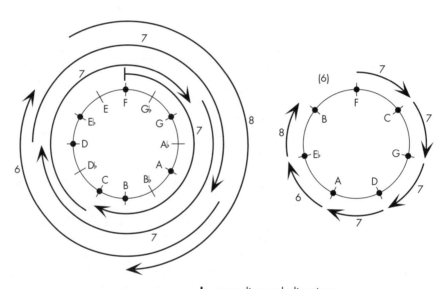

b. ascending melodic minor

The Bisector and the Octatonic Scale

Similarly to the harmonic and ascending melodic minor collections, the octatonic scale, though it is maximally even, cannot be generated by any c distance: It is not a generated collection and, consequently, is not well formed. Furthermore, any collection with a greatest common divisor of crossing lines and dots that is greater than 1 *and* less than the number of dots—or $1 < \text{GCD (c, d)} < d$—*has no generator*. The octatonic scale has no generator, because GCD (12, 8) = 4, and $1 < 4 < 8$ (thus, no generator). However, an approximate bisector of this collection, 5 (or 3), can be used to produce this scale, as shown in Figure 2.4. Again, we obtain this result because these approximate bisectors are coprime with the number of notes in the collection: GCD (5, 8) = 1 and GCD (3, 8) = 1. Of course, the exact bisector is 4, but this bisector will not produce the collection: GCD (4, 8) = 4 (not 1, thus not coprime). However, Jay Rahn accepts any division of the octave between one-third and two-thirds of the total as an *approximate* half; therefore, the bisector 5 (or 3) may be used to produce the collection in lieu of a generator.[13]

Because the c distances that correspond to the constant d distance of 5 alternate between 8 and 7, it is clear that this collection is not generated. Nevertheless, in order to carry out a similar procedure for the octatonic collection as for the other collections explored, Exercise 2.12 employs a circle that is labeled based on this bisector of the octatonic collection. This strategy approximates the circle-of-fifth configurations used for the diatonic, pentatonic, harmonic minor, and ascending melodic minor collections. However, in the octatonic collection these "fifths" are alternately perfect fifths (with c distances of 7) and minor sixths or augmented fifths (with c distances of 8) in terms of traditional interval identification.

Figure 2.4 The octatonic collection constructed by its bisector (5). The c distances alternate between 8 and 7, approximating "fifths." Although this collection is not generated, because the c distances are not fixed, the d distances are constant.

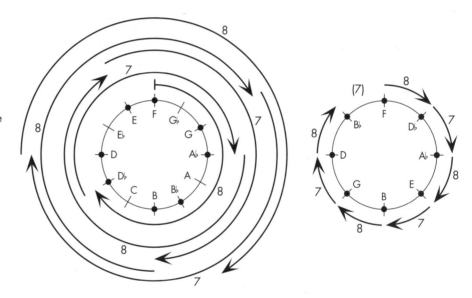

Testing Collections with Bisectors

Complete Exercise 2.12, using the provided circle diagrams and following the procedures used in Exercises 2.9 and 2.11. After you have completed your work, determine if structure implies multiplicity for each collection. Then, compare your answers with those provided in Solution 2.12.

Using Exercises 2.6 and 2.7 from earlier in the chapter, plot the initial series on the circle diagrams and complete the tables as indicated.

Group of series from exercise	Initial series of notes	Drawn from which collection	Distances between dots around circle of fifths
2.6a (p. 69)			

Interval pattern	Number of occurrences

Group of series from exercise	Initial series of notes	Drawn from which collection	Distances between dots around circle of fifths
2.6b			

Interval pattern	Number of occurrences

Group of series from exercise	Initial series of notes	Drawn from which collection	Distances between dots around circle of "fifths"
2.7c (p. 75)			

Interval pattern	Number of occurrences

Does structure imply multiplicity for: the harmonic minor scale?_____; the ascending melodic minor scale? _____; the octatonic scale _____.

In each case, as you likely were able to see immediately, structure cannot imply multiplicity because cardinality does not equal variety. The two properties go hand in hand; the intervals between notes around the circles cannot match the number of times that different interval patterns can be formed by transpositions of series unless cardinality equals variety. For example, the series drawn from the harmonic minor scale, first explored in Exercise 2.6a, projects seven different interval patterns, each occurring once. On the other hand, the number of notes in the series suggests that only four should be formed, and the distances between adjacent dots suggest how many of each of these four patterns should occur. This discrepancy cannot be rectified; therefore, structure does not imply multiplicity for series of notes drawn from the harmonic minor scale. The ascending melodic minor scale, first explored in Exercise 2.6b, presents the same problem because there are five different interval patterns (2, 1, 2, 1, 1) rather than the expected four (2, 2, 2, 1), based on the corresponding circle shown in Solution 2.12. Similarly, structure does not imply multiplicity for the octatonic scale, taken originally from Exercise 2.7c and depicted in the last table of Solution 2.12.

Although we used bisectors instead of generators for the diagrams of the harmonic minor, ascending melodic minor, and octatonic collections (because these collections, again, are not generated), the real problem is the number of notes compared with the number of interval patterns formed—cardinality does not equal variety. We could try some other substitutes for the generator, other than the bisector, but the results would be the same because the number of notes and the number of interval patterns formed are different.

The Generated Whole-Tone Collection

One important feature that the diatonic and pentatonic collections share is that both are generated, as established earlier. However, whereas all collections with cardinality equals variety and structure implies multiplicity are in fact generated, not all generated collections have these properties. For example, the whole-tone collection is generated (as well as maximally even and well formed). In this case, the generator equals the whole steps between consecutive notes that form the scale.[14] For any generated collection whose greatest common divisor of crossing lines and dots is equivalent to the number of dots, the generator (g) is always c divided by d (and c − g). Or, in formulaic terms, if GCD of (c, d) = d, then $\frac{c}{d}$ = g. Thus, for the whole tone collection, GCD (12, 6) = 6, and $\frac{12}{6}$ = 2. Consequently, the whole-tone collection can be generated by a c distance of 2, as shown in Figure 2.5, or a c distance of 10: g and c − g, or 2 and 12 − 2. These generators correspond to ascending and descending whole steps (d distances of 1 and 5). Thus, the resulting circle, shown on the right in Figure 2.5, looks essentially the same as

The initial series plotted on the circle diagram and the completed table for each indicated exercise from earlier in the chapter

Group of series from exercise	Initial series of notes	Drawn from which collection	Distances between dots around circle of fifths
2.6a	D–E–F–G	D harmonic minor	2, 2, 2, 1

Interval pattern	Number of occurrences
2–1–2–7	1
1–2–2–7	1
2–2–1–7	1
2–1–3–6	1
1–3–1–7	1
3–1–2–6	1
1–2–1–8	1

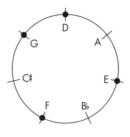

Group of series from exercise	Initial series of notes	Drawn from which collection	Distances between dots around circle of fifths
2.6b	E–F♯–G–A	E ascending melodic minor	2, 2, 2, 1

Interval pattern	Number of occurrences
2–1–2–7	2
1–2–2–7	1
2–2–2–6	2
2–2–1–7	1
1–2–1–8	1

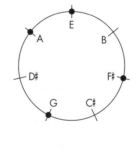

Group of series from exercise	Initial series of notes	Drawn from which collection	Distances between dots around circle of "fifths"
2.7c	C–D–E♭	Octatonic	2, 3, 3

Interval pattern	Number of occurrences
2–1–9	4
1–2–9	4

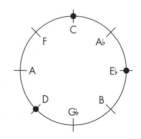

the version with twelve crossing lines, except that every other line is removed. Despite the presence of a generator in this maximally even and well-formed collection, because there is only one interval pattern in each group of transposed series (as shown in Exercise 2.7b), this single interval pattern will appear six times. Therefore, structure does not imply multiplicity.

Figure 2.5 The generated whole-tone collection

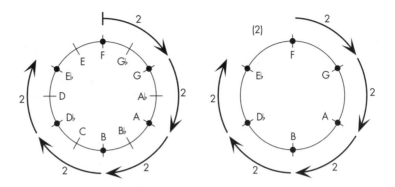

*S*UMMARY AND EXTENSIONS

At this point we can safely conclude that cardinality equals variety and structure implies multiplicity are not properties of all maximally even collections. Although the diatonic and pentatonic collections are maximally even and exhibit both cardinality equals variety and structure implies multiplicity, the other maximally even collections that we tested (the whole tone and octatonic scales) do not exhibit these properties. Furthermore, cardinality equals variety and structure implies multiplicity are not properties of all generated collections. The whole-tone collection is generated, but the single distinct pattern that will be formed when any series is transposed within this collection will not even remotely resemble the numbers of patterns suggested by dots placed around the circle.

Clearly, something else is at work in producing cardinality equals variety and structure implies multiplicity. Given all that we have learned about various collections so far in this book, what aspects of the pentatonic and diatonic collections set them apart from these other collections? Why are these the only collections we studied that display both of these properties?

Myhill's Property

One important similarity between the pentatonic and diatonic collections that we have not mentioned is their comparable intervallic configurations. Return to the tables of intervals in Figure 1.9 (p. 30), Exercise 1.6d (p. 32), and Exercise 1.7c (p. 36). According to these tables, the pentatonic and diatonic collection each have exactly two interval qualities (c distances) for every interval type (d distance). None of the other scales that we have studied have exactly two c distances for every d distance; some have an inconsistent number of distances, others have more or less than the two exhibited by the pentatonic and diatonic collections. Any collection that has *exactly* two c distances for every d distance will exhibit cardinality equals variety and structure implies multiplicity. This special attribute is called *Myhill's property*.[15] Any collection that has Myhill's property (exactly two c distances for every d distance) will exhibit both of these special properties, regardless of the number of notes in the series or even the number of notes in the chromatic universe (such as microtonal divisions of the octave). Furthermore, any collection with Myhill's property will be well formed and generated.[16]

Therefore, the comparable intervallic configurations of the pentatonic and diatonic collections that we explored in Chapter 1 are directly related to cardinality equals variety and structure implies multiplicity. Chapter 2 has provided you with an opportunity to observe these two important aspects of these collections independently in the interval patterns formed by various series of notes drawn from diatonic and pentatonic scales.

The Source of Cardinality Equals Variety and Structure Implies Multiplicity

Why does cardinality equal variety and structure imply multiplicity for collections with Myhill's property, but not for other collections? The answer to this intriguing question is suggested by the last circle diagram of Figure 2.1, repeated here as Figure 2.6a. The key to the solution is the single interval that completes the circle. As stated earlier in this chapter, for collections that are maximally even and well formed, there never will be more than *one* c distance corresponding to the d distance of the generator that does not match the generator (g). Furthermore, this single c distance will always be g $\pm$ 1 (the generating c distance plus or minus one) if c and d are coprime, or GCD of (c, d) = 1. Thus, in the diatonic collection this single interval—with a c distance of 6, one less than the generator—is partially the result of the fact that the diatonic collection is coprime. Because the diatonic collection is generated, all of the other c distances between adjacent lines around the circle are equivalent (7). This basic configuration is essential to cardinality equals variety and structure implies multiplicity.

If, for example, we plot a series of three notes around the circle—such as F, G, and D, as shown in the first circle diagram of Figure 2.6b (top left)—there are only three distinct locations, with respect to the dots, where the unique interval in the circle (6) could possibly be located as the series is rotated (or transposed) around the circle. The interval 6 could appear between the dots that are four lines apart, as shown in the four circles of Figure 2.6b; the interval 6 could appear between the dots that are one line apart, as shown in Figure 2.6c; or it could appear between the dots that are two lines apart, as shown in the two circles of Figure 2.6d. Because the location of this unique interval will directly affect the total number of half steps between the different pairs of dots, the three-note series can form only three distinct interval patterns (Figure 2.6b, Figure 2.6c, and Figure 2.6d), or cardinality equals variety.[17] Furthermore, the unique interval (6) eventually will appear once in each of the seven possible locations, as the dots are rotated around the circle. As Figure 2.6b illustrates, there are four possible locations where the interval 6 may appear between the dots that are four lines apart, because, in effect, there are four "spaces" between these two dots available for the placement of this unique interval. Similarly, as Figure 2.6c shows, there is only one possible location for the interval 6 to appear between the dots that are only one line apart. In addition, as seen in Figure 2.6d, there are exactly two possible locations where the interval 6 may appear between the dots that are two lines apart. Thus, the distances between dots initially placed on F–G–D (4, 1, and 2) indicate exactly how many of each interval pattern that can be formed, or structure implies multiplicity.

Figure 2.6 The intervallic structure of the generated diatonic collection (a) and its relationship to cardinality equals variety and structure implies multiplicity. All seven rotations of the series F–G–D appear in b through d.

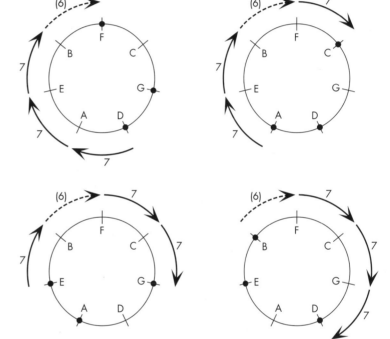

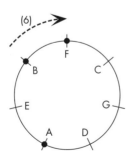

a. the generated diatonic collection

b. four patterns with the interval 6 appearing between dots that are four lines apart

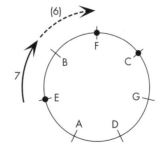

c. one pattern with the interval 6 appearing between dots that are one line apart

d. two patterns with the interval 6 appearing between dots that are two lines apart

You may wish to verify for yourself that choosing any other initial pattern of dots within the diatonic collection produces the same results. Also, you can apply the same set of procedures to the pentatonic collection, (using the generator to build the collection and rotating a pattern of dots around the circle, as in Figure 2.6); however, the interval remaining to complete the circle is 8 (still g ± 1). Interestingly enough, although cardinality does not equal variety and structure does not imply multiplicity for the generated whole-tone collection, if you take any one note away from this collection, the results change dramatically. The collection of five remaining notes is a generated collection in which cardinality equals variety and structure implies multiplicity. Can you explain why? (Hint: Follow the procedure demonstrated in Figure 2.6, but use a five-dot circle diagram generated by whole tones. Also, you may want to write out a group of transposed series, such as those shown in Exercise 2.7b [on p. 74], to observe this phenomenon firsthand.) This intriguing question is left for you to explore on your own as you wish.

A Look Forward

In the final chapter of this book, we will concentrate solely on the diatonic collection in an attempt to determine more of its attributes. Although many of the characteristics that we will observe in the diatonic collection may also occur in the pentatonic collection due to the similarities between the two collections that we have already established, we will confine our discussion to the more common and familiar diatonic collection. By limiting our focus in this way, we hope to provide more information on important structural aspects of this remarkable collection of notes, the collection that is likely to be the primary foundation of your studies in tonal music theory.

TRIADS AND SEVENTH CHORDS AND THEIR STRUCTURES

3

*F*ROM COLLECTION TO CHORD

In the first two chapters of this book, we have studied some important properties associated with the diatonic collection. We have observed that the diatonic collection is the only seven-note collection that is maximally even, and we have learned that cardinality equals variety and structure implies multiplicity for any series of notes drawn from the diatonic collection. Furthermore, we have observed that the diatonic collection is a generated, well-formed, and deep scale with Myhill's property. Based on our observations of all of these properties, we have seen that the diatonic collection is a very special group of notes, which in part may suggest why composers have been drawn to this particular collection for such a long time. The diatonic collection serves as the primary basis of much tonal music; therefore, it seems prudent to determine as much as we can about its structure, as well as the structures of its constituent parts.

In this chapter we will explore diatonic triads and seventh chords, the most common sonorities that are drawn from the diatonic collection in tonal music. We will attempt to determine which of the primary properties discussed in this text apply to these musical constructs, and we will try to relate our earlier observations about the diatonic collection as a whole to these essential chords. First, we will examine diatonic triads and seventh chords to determine if these structures are maximally even. As we have previously concluded, the fact that the diatonic collection itself is maximally even is a fundamental property of this important collection; however, we have yet to examine the most important musical components drawn from this collection, diatonic triads and seventh chords, to see if they also exhibit this property. Finally, we will explore the ideas of cardinality equals variety and structure implies multiplicity in connection with diatonic triads and seventh chords. Exploring these two properties, which are also fundamental characteristics of the diatonic collection, in association with diatonic triads and seventh chords will reveal important attributes of the configuration and structure of these chords.

MAXIMALLY EVEN TRIADS AND SEVENTH CHORDS

Diatonic Chords and New Circle Diagrams

In this exploration of diatonic triads and seventh chords, we shall combine some of the techniques employed in both of the previous chapters. In Chapter 1 we used a circle with twelve crossing lines to represent the chromatic scale, and we labeled the crossing lines using this ascending scale in a clockwise manner around the circle. We also learned how to ascertain whether a circle diagram is maximally even by determining the distances between every pair of dots around the circle, both in terms of the number of dots (d distances) and the number of crossing lines (c distances) spanning each pair of dots. In Chapter 2 we used a circle with seven lines to represent the diatonic collection, and in connection with our study of structure implies multiplicity, we used the generator to label the crossing lines around the circle (corresponding to the circle of fifths). In the present chapter, by combining aspects of the techniques used in both of these earlier chapters, we can explore maximal evenness of triads and seventh chords with respect to the diatonic collection.

Exercise 3.1a presents two circle diagrams each with seven crossing lines to represent the diatonic collection. First, arrange three and four dots around the circles so that the dots are maximally even, or spread out as much as possible. Let your experience from Chapter 1 guide you in placing the dots around the circles; you need not use the interval definition as an aid at this point, though we will return to this definition in subsequent exercises. Next, instead of labeling the crossing lines using the generator, as in Chapter 2, assign note names to the seven lines using the *stepwise* notes of the D ascending major scale moving clockwise around the circle. In this way, your placement of dots around the circle will imply that the associated musical structure formed by the dots will be maximally even, because the concept of maximal evenness depends on the intervallic relationships between notes measured in consecutive steps, not by means of the circle of fifths. Finally, identify the precise musical structures formed by the dots you placed around the circles. Before checking your solution to this exercise, repeat the process in Exercises 3.1b, 3.1c, and 3.1d. However, in each exercise, keep the same placements of dots as in Exercise 3.1a, but alter your assignment of notes so that the notes of the D major scale correspond to different lines around the circle—still moving in a clockwise manner and in a scalar fashion but starting the scale on a different crossing line each time.

EXERCISE 3.1a Arrange three and four dots around the circles (of seven lines) below so that the dots are maximally even. Assign note names to the seven lines using the stepwise notes of the D ascending major scale moving clockwise around the circles. Then, answer the question that follows.

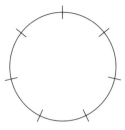 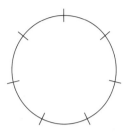

What precise musical structure is formed?

3 dots _____ 4 dots _____

EXERCISE 3.1b Repeat the arrangement of three and four dots that you used in Exercise 3.1a on the following circles. Assign different note names to each of the seven lines (beginning at a different place in the circle, but still using the stepwise notes of the same scale in a clockwise manner).

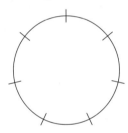 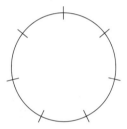

What precise musical structure is formed?

3 dots _____ 4 dots _____

EXERCISE 3.1c Again, repeat the arrangement of three and four dots that you used previously on the following circles. Assign different note names to each of the seven lines (beginning at a different place in the circle, but still using the stepwise notes of the same scale in a clockwise manner).

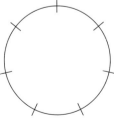 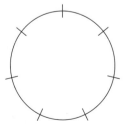

What precise musical structure is formed?

3 dots _____ 4 dots _____

Once more, repeat the arrangement of three and four dots that you used in the other parts of the exercise on the following circles. Assign different note names to each of the seven lines (beginning at a different place in the circle, but still using the stepwise notes of the same scale in a clockwise manner).

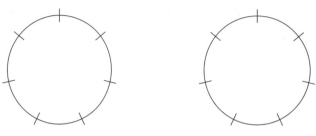

What precise musical structure is formed?

3 dots _____ 4 dots _____

A Maximally Even Hypothesis

Delay checking your solutions and look over your work in this exercise. Observe any similarities or differences among the various chords you have constructed. Then in Exercise 3.2, make a generalized statement about what you have observed in the circles and what you have learned about maximal evenness, beyond what you already knew from your work in Chapter 1. Be sure to mention the kind of collection from which the musical structures are drawn. Try to be as general as possible; consider whether your answer would be different if you had used an E♭ or A major scale instead of, or in addition to, the D major scale that you used.

Make a generalized statement about what you have observed in the circles in Exercise 3.1 and what you have learned about maximally even structures by completing that exercise.

_____

_____

_____

Calculating Dot Placements and Diatonic Circles

There are numerous possible solutions to Exercise 3.1, depending on how you labeled the lines around the circles in the four parts of the exercise and on where you placed the dots around the circles. The number of distinct ways to assign lines around a circle may be determined by the method introduced in Chapter 1. Figure 3.1 calculates the number of possible ways to arrange various numbers of dots around a circle with seven lines according to the formula. Figures 3.1a and 3.1b indicate the number of possible ways to place three and four dots around a circle with seven lines, as directed in Exercise 3.1. The values used in this formula are substantially different than the values we explored in connection with twelve-line circles. Because of the fact that c is now 7 rather than 12, almost any number of dots placed around a circle may be arranged in seven different ways. The greatest common divisor of seven (c) and any number of dots less than seven (d) will always be one, because seven is a prime number. The only exception, shown in Figure 3.1c, is that there is, of course, only one way to place all seven dots around a circle with seven lines—one dot on each line.

Figure 3.1 The number of possible ways to arrange various numbers of dots around a circle with seven lines

a. 3 dots $\quad \dfrac{c}{GCD(c,\,d)} = \dfrac{7}{GCD(7,\,3)} = \dfrac{7}{1} = \boxed{7}\quad$ correct solutions

b. 4 dots $\quad \dfrac{c}{GCD(c,\,d)} = \dfrac{7}{GCD(7,\,4)} = \dfrac{7}{1} = \boxed{7}\quad$ correct solutions

c. 7 dots $\quad \dfrac{c}{GCD(c,\,d)} = \dfrac{7}{GCD(7,\,7)} = \dfrac{7}{7} = \boxed{1}\quad$ correct solutions

Labeling Lines and Placing Dots

Solution 3.1 breaks down the different components of the assigned tasks to aid you in evaluating your own work. First, Solution 3.1a shows all possible ways to assign lines around a circle using the ascending D major scale in a clockwise arrangement, as directed. Each of the circles in Exercise 3.1 (a through d) should correspond exactly with one of the provided circles in Solution 3.1a. Next, Solution 3.1b shows all seven possible maximally even placements of three dots and four dots—with D labeled on the crossing line at the top of the circle, as an example. Regardless of how you labeled your crossing lines, the placement of dots for each of the three-dot and four-dot circles in your solutions to the exercise should exactly match one of these patterns of dots, even if your labels are different. If you labeled any of the circles in Exercise 3.1 starting with D on the line at the top of the circle, the circles will exactly match the provided solutions. Otherwise, you can determine if your solutions are correct by comparing your work with these provided diagrams and rotating the line labels as necessary to orient these solutions with your own work. Finally, Solution 3.1c shows all possible precise musical structures that may be formed by the notes corresponding to the dots in each circle. All of your answers to this part of Exercise 3.1a–d should appear directly and precisely (but in any order) on these lists, depending on your placement of dots and how you labeled the lines. However, you will not have arrived at all of these formations, because you were instructed only to complete four circles.

All possible ways to assign lines using the D major scale, as directed. (This solution applies to all parts of Exercise 3.1, not just to 3.1a.)

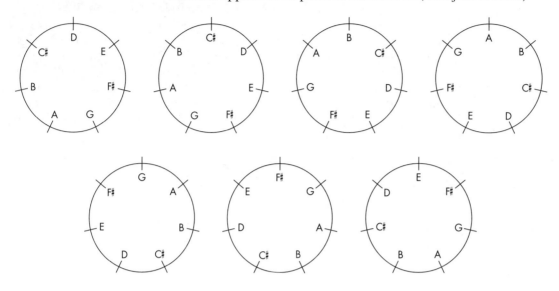

All possible maximally even placements of three and four dots, with D labeled at the top of the circle. (This solution applies to all parts of Exercise 3.1, not just to 3.1b.)

3 dots

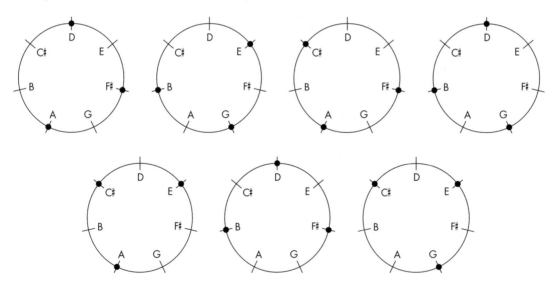

4 dots

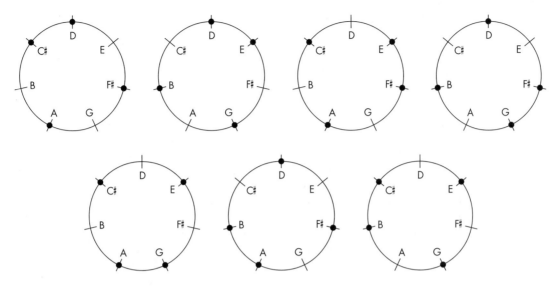

⌇⌇

SOLUTION 3.1c   The precise musical structures that may be formed for each circle by the notes corresponding to the indicated number of dots. (This solution applies to all parts of Exercise 3.1, not just to 3.1c.)

3 dots _____ D major, E minor, F♯ minor, G major, A major, B minor, and
_____ C♯ diminished triads

4 dots _____ D major seventh, E minor seventh, F♯ minor seventh,
_____ G major seventh, A dominant seventh, B minor seventh,
_____ and C♯ half-diminished seventh chords

⌇⌇

Second-Order Maximal Evenness

As shown in Solution 3.2, your generalized statement about what you observed in these circles should acknowledge that diatonic triads and seventh chords are maximally even with respect to the diatonic collection. We had already determined that the diatonic collection is itself maximally even, based on our work in Chapter 1, and now we have established that diatonic triads and seventh chords are maximally even with respect to it. Thus, there is a nested maximal evenness extending from the diatonic triad or diatonic seventh chord up to the chromatic universe by means of the intervening diatonic collection. A structure that is maximally even in this way (by means of another maximally even collection that contains it) is termed *second-order maximally even*.

A generalized statement about the circles in Exercise 3.1 and maximally even structures.

> Triads and seventh chords formed from a diatonic collection are
>
> maximally even with respect to the diatonic collection.

〜

Testing Triads for Maximal Evenness

In Exercises 3.3 and 3.4 we will attempt to confirm our generalized statement about triads and seventh chords by using the more formal definition of maximal evenness, based on intervals, and in so doing we will summarize what we have learned about triads and seventh chords so far in this book. Exercise 3.3 shows several circle diagrams to complete. These diagrams are arranged in pairs; in each pair there is one seven-line circle representing the diatonic collection corresponding to the E major scale and one twelve-line circle representing the chromatic universe. On each pair of circles, plot an example of the indicated triad quality, using the same triad for each circle in the pair. Test each triad to see if it is maximally even according to the intervallically based definition, in reference to both the diatonic collection ($c = 7$) and the chromatic universe ($c = 12$). In each case, a triad is maximally even if for each d distance (in dots) there are only one or two possible c distances (in crossing lines), and if there are two c distances for a particular d distance, then the c distances are consecutive numbers. From a musical perspective, the c distances on the circles with twelve lines, representing the chromatic universe, correspond to half steps (as we observed in Chapter 1), whereas the c distances on the circles with seven lines, representing the diatonic collection, correspond to diatonic scale steps. Thus, in counting c distances on the diatonic circles, you are counting generic scale steps, regardless of whether the steps are half steps or whole steps; in counting c distances on the chromatic circles, you are counting specific half steps. In both types of circle diagrams, d distances are counted in the same way; to determine the d distance between two dots, count the number of dots from one dot to the other (but excluding the initial dot). Complete the provided tables by indicating the distances between each pair of dots. If any of the triad qualities cannot be plotted using the given notes around a circle, place a question mark in the center of its circle diagram, leave the table blank, then answer the question.

EXERCISE 3.3 Plot an example of each of the following triads on the provided circles, if possible. Test each plotted triad to see if it is maximally even according to the interval definition, based on the diatonic collection (the circles on the left) and the chromatic universe (the circles on the right).

a. Major triad

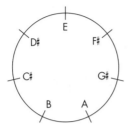

CLOCKWISE DISTANCE BETWEEN DOTS	
d distance	**c distance**

Maximally even?_____

CLOCKWISE DISTANCE BETWEEN DOTS	
d distance	**c distance**

Maximally even?_____

b. Minor triad

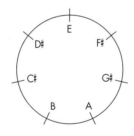

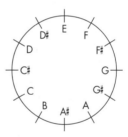

CLOCKWISE DISTANCE BETWEEN DOTS	
d distance	**c distance**

Maximally even?_____

CLOCKWISE DISTANCE BETWEEN DOTS	
d distance	**c distance**

Maximally even?_____

c. Diminished triad

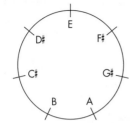 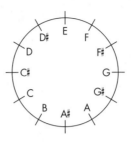

CLOCKWISE DISTANCE BETWEEN DOTS	
d distance	**c distance**

Maximally even?_____

CLOCKWISE DISTANCE BETWEEN DOTS	
d distance	**c distance**

Maximally even?_____

d. Augmented triad

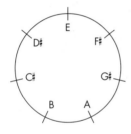 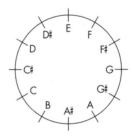

CLOCKWISE DISTANCE BETWEEN DOTS	
d distance	**c distance**

Maximally even?_____

CLOCKWISE DISTANCE BETWEEN DOTS	
d distance	**c distance**

Maximally even?_____

Maximally Even Triads

There are several possibilities for the placement of dots around the circles in this exercise—except for the diminished triad, which has only one possible configuration, and the augmented triad, which cannot be plotted on the diatonic circle. Solution 3.3 shows only one possible arrangement of dots for each circle as an example. However, regardless of which particular triads you chose to plot, your tables and answers to the questions should match those provided in Solution 3.3. Major, minor, and diminished triads are all maximally even with respect to the diatonic collection and *not* maximally even with respect to the chromatic universe (as shown in Solutions 3.3a, b, and c). They are second-order maximally even. On the other hand, the augmented triad is not maximally even with respect to the diatonic collection—it cannot even be formed from the notes of the diatonic collection—but it *is* maximally even with respect to the chromatic universe (as shown in Solution 3.3d).

SOLUTION 3.3   One possible way to plot an example of each of the following triads; other solutions are also possible. Also, a test to determine if each triad is maximally even, based on the diatonic collection (the circles on the left) and the chromatic universe (the circles on the right)

a. Major triad

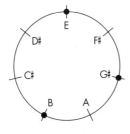
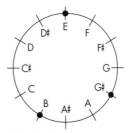

CLOCKWISE DISTANCE BETWEEN DOTS	
d distance	c distance
1	2, 3
2	4, 5
Maximally even?	Yes

CLOCKWISE DISTANCE BETWEEN DOTS	
d distance	c distance
1	3, 4, 5
2	7, 8, 9
Maximally even?	No

b. Minor triad

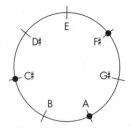

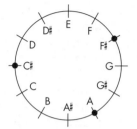

CLOCKWISE DISTANCE BETWEEN DOTS	
d distance	**c distance**
1	2, 3
2	4, 5
Maximally even?	Yes

CLOCKWISE DISTANCE BETWEEN DOTS	
d distance	**c distance**
1	3, 4, 5
2	7, 8, 9
Maximally even?	No

c. Diminished triad

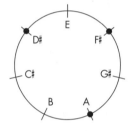

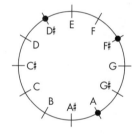

CLOCKWISE DISTANCE BETWEEN DOTS	
d distance	**c distance**
1	2, 3
2	4, 5
Maximally even?	Yes

CLOCKWISE DISTANCE BETWEEN DOTS	
d distance	**c distance**
1	3, 6
2	6, 9
Maximally even?	No

d. Augmented triad

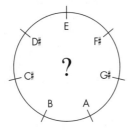

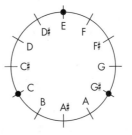

CLOCKWISE DISTANCE BETWEEN DOTS	
d distance	**c distance**
Maximally even?	No

CLOCKWISE DISTANCE BETWEEN DOTS	
d distance	**c distance**
1	4
2	8
Maximally even?	Yes

Interval Content of Triads

Let us take a moment to examine these tables more closely to see if we can draw any other conclusions about relationships among diatonic triads. First, notice that the tables you completed for the major and minor triads (Exercises 3.3a and 3.3b) are identical to each other in both the diatonic and the chromatic cases. Also, the diminished triad (Exercise 3.3c) has the same values in the table for the diatonic circle as the major and minor triads do, but it has different values in the table for the chromatic circle than those of the major and minor triads. Clearly, these three triads sound different; play various examples of these triads on the piano to verify this difference for yourself. Yet, despite these differences in sound, these three triads have exactly the same interval content in reference to the diatonic collection, as we have shown in the tables. Why does this correspondence among these different triads appear in the diatonic tables? The diatonic tables match because all of these triads are drawn from the diatonic collection. Thus, in all cases the distances between adjacent notes are either thirds or fourths, and the distances between the other pairs of notes—represented by dots that are separated from each other by another dot, clockwise—are either fifths or sixths, in the traditional musical sense of those intervals. The tables corresponding to the diatonic circles indicate nothing about the *qualities* of those intervals because we are only counting scale steps, not half steps, and the qualities of the intervals surely impact how the triads will sound.

On the other hand, the correspondence between the major and minor triads in the chromatic tables (Exercises 3.3a and 3.3b) is perhaps more surprising. In this case, we *are* counting half steps; therefore, these triads must be even more closely related because they contain exactly the same intervals, in terms of both type and quality, though they certainly do not sound the same. This aspect of the direct relationship between these two chords can be understood, in the traditional musical sense, by considering their interval contents. The major triad is built from the root up with a major third, then a minor third, and with a perfect fourth to complete the octave. Conversely, the minor triad is built from the root up with a minor third, then a major third, and with a perfect fourth to complete the octave. The two triads, therefore, *contain* the same intervals, though the *placement* of these intervals makes all the difference in the resulting sound.

Inversionally Related Triads

Figure 3.2 shows another, even stronger, aspect of the relationship between major and minor triads. In addition to the fact that the two triads have the same intervallic content, major and minor triads are also inversionally related.[1] The dots corresponding to a major triad (taken from Solution 3.3a, and shown as the first circle diagram in the figure) are inverted (or flipped) around a dashed line drawn vertically through the center of the first circle diagram to produce the second circle diagram. Notice that the E at the top of the first circle diagram stays in the same place after inversion around the dashed line, as shown in the second circle diagram. However, the B moves to A, and the G♯ moves to C as a result of this inversion. This new pattern of dots can then be rotated (in this case, counterclockwise by a distance of three lines, or, equivalently, clockwise by a distance of nine lines) to produce the third circle diagram. Rotation of a circle diagram effectively amounts to transposition in musical terms. The inverted (second) circle diagram displays an A minor triad, and transposing this inverted triad down three half steps (or up nine half steps) produces the F♯ minor triad (taken from Solution 3.3b), shown at the right of the figure.[2] Thus, the E major triad and F♯ minor triad shown are inversionally related. And more generally, all major and minor triads are inversionally related.[3] Rotating the inverted diagram by other distances would produce other minor triads. If you wish, you can also verify on your own that inverting a minor triad will produce a major triad by following the process shown in Figure 3.2.

Figure 3.2 Demonstration of inversion for major and minor triads. The major and minor triads shown here are the same as those in Solution 3.3; however, any pair of major and minor triads are inversionally related in the same way.

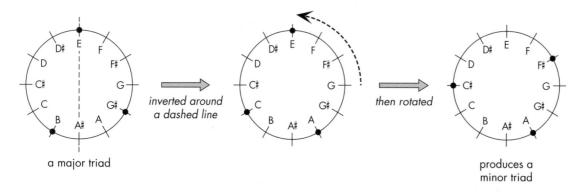

a major triad   inverted around a dashed line   then rotated   produces a minor triad

Testing Seventh Chords for Maximal Evenness

Exercise 3.4 requires the same tasks as in Exercise 3.3 to be performed for seventh chords. Plot the indicated seventh chords on the circles, using the same seventh chord for each circle in the pair. As before, if any of the seventh-chord qualities cannot be plotted using the given notes around a circle, place a question mark in the center of its circle diagram, leave the table blank, then answer the question. In this exercise, for the sake of variety, the diatonic collection used corresponds to the B♭ major scale, rather than the E major scale used in the previous exercise. Remember to count scale steps when using the interval definition of maximally even for the c distances in the diatonic circles, and to count half steps for the c distances in the chromatic circles.

EXERCISE 3.4   Plot an example of each of the following seventh chords on the provided circles, if possible. Test each plotted seventh chord to see if it is maximally even according to the interval definition, based on the diatonic collection (the circles on the left) and the chromatic universe (the circles on the right).

a. Major seventh chord

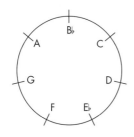

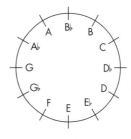

CLOCKWISE DISTANCE BETWEEN DOTS	
d distance	**c distance**

Maximally even?_____

CLOCKWISE DISTANCE BETWEEN DOTS	
d distance	**c distance**

Maximally even?_____

b. Minor seventh chord

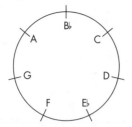

CLOCKWISE DISTANCE BETWEEN DOTS	
d distance	**c distance**

Maximally even?_____

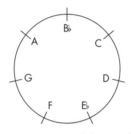

CLOCKWISE DISTANCE BETWEEN DOTS	
d distance	**c distance**

Maximally even?_____

c. Dominant seventh chord

CLOCKWISE DISTANCE BETWEEN DOTS	
d distance	**c distance**

Maximally even?_____

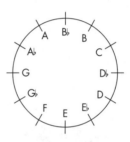

CLOCKWISE DISTANCE BETWEEN DOTS	
d distance	**c distance**

Maximally even?_____

d. Half-diminished seventh chord

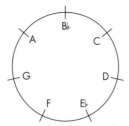

CLOCKWISE DISTANCE BETWEEN DOTS	
d distance	**c distance**

Maximally even?_____

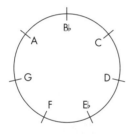

CLOCKWISE DISTANCE BETWEEN DOTS	
d distance	**c distance**

Maximally even?_____

e. Diminished seventh chord

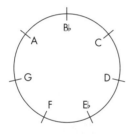

CLOCKWISE DISTANCE BETWEEN DOTS	
d distance	**c distance**

Maximally even?_____

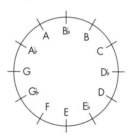

CLOCKWISE DISTANCE BETWEEN DOTS	
d distance	**c distance**

Maximally even?_____

Maximally Even Seventh Chords

As in the previous exercise, there are several possibilities for the placement of dots around the circles—except for the dominant seventh and half-diminished seventh chords (Solutions 3.4c and 3.4d), each of which has only one possible configuration; and the diminished seventh chord (Solution 3.4e), which cannot be plotted on the diatonic circle. Solution 3.4 shows one possible arrangement of dots for each circle. Regardless of which particular seventh chords you chose to plot, your tables and answers to the questions should match those provided in Solution 3.4. Major, minor, dominant, and half-diminished seventh chords (shown in Solutions 3.4a, b, c, and d) are all maximally even with respect to the diatonic collection and *not* maximally even with respect to the chromatic universe. They are second-order maximally even. On the other hand, the diminished seventh chord (shown in Solution 3.4e) is not maximally even with respect to the diatonic collection—it cannot even be formed from the notes of the diatonic collection—but it *is* maximally even with respect to the chromatic universe.

SOLUTION 3.4 One possible way to plot an example of each of the following seventh chords; other solutions are also possible. Also, a test to determine if each seventh chord is maximally even, based on the diatonic collection (the circles on the left) or the chromatic universe (the circles on the right)

a. Major seventh chord

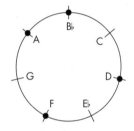 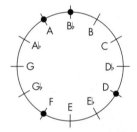

CLOCKWISE DISTANCE BETWEEN DOTS	
d distance	**c distance**
1	1, 2
2	3, 4
3	5, 6
Maximally even?	Yes

CLOCKWISE DISTANCE BETWEEN DOTS	
d distance	**c distance**
1	1, 3, 4
2	5, 7
3	8, 9, 11
Maximally even?	No

b. Minor seventh chord

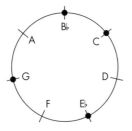

CLOCKWISE DISTANCE BETWEEN DOTS	
d distance	**c distance**
1	1, 2
2	3, 4
3	5, 6
Maximally even?	Yes

CLOCKWISE DISTANCE BETWEEN DOTS	
d distance	**c distance**
1	2, 3, 4
2	5, 7
3	8, 9, 10
Maximally even?	No

c. Dominant seventh chord

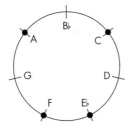

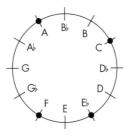

CLOCKWISE DISTANCE BETWEEN DOTS	
d distance	**c distance**
1	1, 2
2	3, 4
3	5, 6
Maximally even?	Yes

CLOCKWISE DISTANCE BETWEEN DOTS	
d distance	**c distance**
1	2, 3, 4
2	5, 6, 7
3	8, 9, 10
Maximally even?	No

d. Half-diminished seventh chord

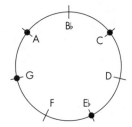

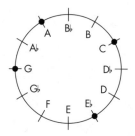

CLOCKWISE DISTANCE BETWEEN DOTS	
d distance	**c distance**
1	1, 2
2	3, 4
3	5, 6

Maximally even? _____ Yes _____

CLOCKWISE DISTANCE BETWEEN DOTS	
d distance	**c distance**
1	2, 3, 4
2	5, 6, 7
3	8, 9, 10

Maximally even? _____ No _____

e. Diminished seventh chord

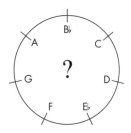

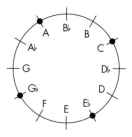

CLOCKWISE DISTANCE BETWEEN DOTS	
d distance	**c distance**

Maximally even? _____ No _____

CLOCKWISE DISTANCE BETWEEN DOTS	
d distance	**c distance**
1	3
2	6
3	9

Maximally even? _____ Yes _____

Interval Content of Seventh Chords

Again let us pause to examine some of the relationships among the chords illustrated by the tables. As we observed regarding the triads in Exercise 3.3, all diatonic seventh chords exhibit identical interval contents with respect to the diatonic collection. By now it probably is clear that this correspondence occurs simply because all of these chords are *seventh chords* drawn from the diatonic collection and therefore are built with stacked thirds.

Perhaps more interesting are the relationships between intervals of the different seventh chords in relation to the chromatic universe. First, contrary to what we observed with triads, the major and minor seventh chords have different interval contents. Although the triad portions of these seventh chords have the same interval contents, as we observed in Exercises 3.3a and 3.3b, the addition of a seventh to each chord changes the interval content significantly. The fact that the major seventh chord (Exercise 3.4a) is built with the interval of a major seventh above the root, whereas the minor seventh chord (Exercise 3.3b) is built with the interval of a minor seventh above the root can be seen in the chromatic tables by comparing the d distances of three dots. The major seventh chord includes the value 11 (for eleven half steps, or a major seventh), whereas the minor seventh chord includes the value 10 (for ten half steps, or a minor seventh). This difference is also seen in the d distance of 1 dot where the major seventh chord includes the interval of one half step, whereas the minor seventh chord includes the interval of two half steps—in each case, from the seventh of the chord up to the root above, an inversion of the interval of a seventh becomes the interval of a second.

On the other hand, the dominant seventh chord (Exercise 3.4c) and the half-diminished seventh chord (Exercise 3.4d) have the same interval contents with respect to the chromatic universe, as exhibited in the tables. Surely, these two chords have distinct sounds; again, verify this difference for yourself by playing several examples of each of these two chord qualities on the piano. However, the two chords are closely related by virtue of their identical interval contents: The dominant seventh chord is built, from the root up, with a major third, then a minor third, another minor third, and a major second to complete the octave; the half-diminished seventh chord, on the other hand, is built with a minor third, another minor third, then a major third, and a major second to complete the octave. Hence, these chords contain the same intervals, but the intervals are arranged in a different order, making all of the difference in the sound of these two chords.

Inversionally Related Seventh Chords

Dominant and half-diminished seventh chords also are inversionally related, as demonstrated in Figure 3.3. The dots corresponding to the dominant seventh chord (taken from Solution 3.3c and shown as the first circle diagram in the figure) are inverted (or flipped) around a dashed line drawn vertically through the center of the first circle to produce the second circle diagram. Notice, at the bottom part of the first circle diagram, that inversion around the dashed line moves the dot on F to E♭ and the dot on E♭ to F, effectively resulting in no change, as shown in the second circle diagram. However, the dot on the A moves to B, and the dot on the C moves to A♭ as a result of this inversion. Next, the pattern of dots resulting from this inversion may be rotated (in this case clockwise by a distance of four lines, or four half steps) to produce the third circle diagram. Thus, the F dominant seventh chord and the A half-diminished seventh chord shown here are inversionally related.[4] Rotating the inverted diagram by other distances would produce other half-diminished seventh chords; therefore, all dominant and half-diminished seventh chords are inversionally related. Again, you may wish to verify that inverting a half-diminished seventh chord will produce a dominant seventh chord by following the process shown in Figure 3.3.

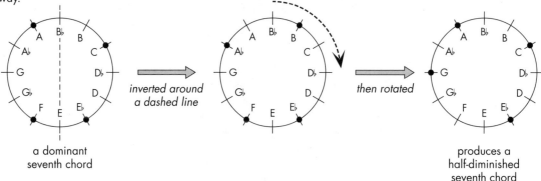

a dominant seventh chord — *inverted around a dashed line* — *then rotated* — produces a half-diminished seventh chord

Maximally Even Triads and Seventh Chords

Exercises 3.3 and 3.4 have confirmed what we have already determined in the first chapter of this book. The only maximally even triad with respect to the twelve-note chromatic universe is the augmented triad, and the only maximally even seventh chord with respect to the chromatic universe is the diminished seventh chord. However, these chords are not drawn from the diatonic collection. Regardless of how they are arranged, the notes of the diatonic collection cannot form augmented triads or diminished seventh chords. Although both of these chords can be drawn from the harmonic minor collection as well as from some other collections, these collections do not exhibit all of the special properties the diatonic collection possesses. By contrast, major, minor, and diminished triads—as well as major, minor, dominant, and half-diminished seventh chords—are all second-order (or nested) maximally even chords by means of the maximally even diatonic collection.

The Augmented Triad in Context

Students often wonder why the augmented triad, built on scale-step 3 in harmonic minor, seems to be given scant attention in tonal theory texts. Exercise 3.3 suggests why the major, minor, and diminished triads might be favored over the augmented triad in diatonic music. Yet the exercise also shows how all of these triads share an important structural element, as each chord is maximally even in a particular context. However, only major, minor, and diminished triads are maximally even in a diatonic context, and these are the chords that appear most often in tonal music. The augmented triad does not appear with frequency in diatonically oriented music; however, the augmented triad began to play a much stronger role as the structural orientation of music shifted toward chromaticism in the later nineteenth century. Your work in Exercise 3.3d corroborates this pattern of usage in that the augmented triad is *not* maximally even in a diatonic context, but it *is* maximally even in a chromatic context.

The Diminished Seventh Chord in Context

On the other hand, the diminished seventh chord plays an increasingly important harmonic role—even in otherwise diatonically oriented music—throughout the development of tonal music in the common-practice period. Although this chord cannot be formed by the notes of the diatonic collection, it is related to those seventh chords that can be formed from the diatonic collection in that all of these seventh chords are maximally even in one way or another. Of course, these observations cannot explain composers' choices regarding harmonic sonorities; many other factors certainly were more directly involved. Nevertheless, our ability to recognize the fundamental properties of the structures of these chords will enhance our understanding of musical structure. Toward this end, the next section of this chapter will explore some additional special properties of diatonic triads and seventh chords.

$\mathcal{V}$ARIETY AND MULTIPLICITY OF DIATONIC CHORDS

In Chapter 2 we learned that any series of notes drawn from the diatonic collection exhibits cardinality equals variety and structure implies multiplicity. In our investigation we studied series of notes that had varying numbers of notes and interval structures. In each case we identified the interval patterns formed by transposing the initial series of notes so that the series began on every note of the diatonic collection. In so doing, we demonstrated that the number of notes in a series, or cardinality, equals the number of distinct interval patterns formed, or variety. Moreover, by observing how each initial series appeared if placed on the circle of fifths (based on the generator), or structure, we were able to ascertain how many times the different interval patterns occurred in each group of transposed series, or multiplicity.

In this section, we will examine triads and seventh chords for these same properties. Whereas in the previous chapter we shaped the series of notes as melodic lines, in this section we will build harmonic sonorities and subject them to the same kind of scrutiny. Through this examination we will have opportunities to learn essential aspects of the variety and multiplicity of triads and seventh chords within the diatonic collection, as well as to review some of the important properties we have been discussing.

Testing Diatonic Chords

Exercises 3.5a and 3.5b contain questions about cardinality equals variety and structure implies multiplicity for diatonic triads and seventh chords, respectively. In order to determine if structure implies multiplicity, the circle diagrams with seven lines are labeled according to the generator (the circle of fifths), as in Chapter 2, rather than in a scalar fashion as in the previous section of this chapter. In addition to asking about the number of different triad and seventh chord qualities that can be formed and the number of individual triads associated with each quality, the exercises also ask you to reveal how you obtained your answers. Rely only on your knowledge of the two properties in determining the answers to these questions and in your specific explanations. At this point, do not work out all of the transpositions, as you did in Chapter 2; we will return to this manual procedure later in conjunction with Roman numerals.

EXERCISE 3.5a

First, plot any triad of your choice using the given notes on the circle diagram (arranged in the circle-of-fifths pattern). Then, answer the questions that follow.

1.

2. Based on cardinality equals variety for triads, how many different triad qualities can be formed in the diatonic collection? _____

3. Explain how you determined the answer to question 2, without writing out all of the triads.

_____

_____

_____

_____

_____

4. Based on structure implies multiplicity (as illustrated by your plotted chord above), how many individual triads will be associated with each quality? (Describe in detail.)

_____

_____

_____

5. Explain how you determined the answer to question 4, without writing out all of the triads.

_____

_____

_____

_____

_____

First, plot any seventh chord of your choice using the given notes on the circle diagram (arranged in the circle-of-fifths pattern). Then, answer the questions that follow.

1.

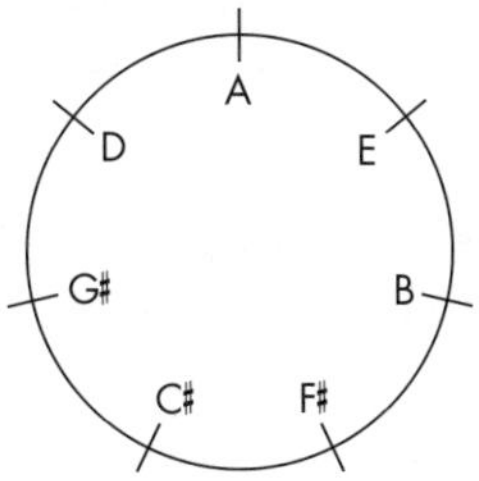

2. Based on cardinality equals variety, how many different seventh-chord qualities can be formed in the diatonic collection? _____

3. Explain how you determined the answer to question 2, without writing out all of the seventh chords.

_____

_____

_____

_____

_____

_____

4. Based on structure implies multiplicity, how many individual seventh chords will be associated with each quality? (Describe in detail.)

_____

_____

_____

_____

5. Explain how you determined the answer to question 4, without writing out all of the seventh chords.

_____

_____

_____

_____

_____

Answers to all of the problems posed in Exercise 3.5 can be determined based on a knowledge of the two properties from Chapter 2. Compare your results with those provided in Solution 3.5. Do not worry about the exact wording of your answers (though the numbers should exactly match, of course); you may have explained how you determined your solutions in a slightly different way. However, in any event your solutions should acknowledge that the number of notes in a triad (three) and seventh chord (four) equals the number of different triad and seventh-chord qualities that can be formed. In addition, you should have indicated that the distances around the circle of fifths determines the number of individual triads or seventh chords that are associated with each quality. The assertions you have made in Exercise 3.5 have involved a more refined application of these two properties and have uncovered a fundamental fact about triads and seventh chords in a diatonic context. We will verify these assertions empirically in Exercise 3.6 to follow.

SOLUTION 3.5a

One possible way to plot a triad using the given notes on the circle diagram (arranged in the circle-of-fifths pattern) and answers to the questions in Exercise 3.5a

1.
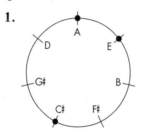

2. Based on cardinality equals variety for triads, how many different triad qualities can be formed in the diatonic collection? ____3____

3. Explain how you determined the answer to question 2, without writing out all of the triads.

    Because there are three notes in a triad, cardinality equals variety

    indicates that there will be exactly three different interval patterns

    formed by transpositions of the triad. Therefore, exactly three

    different triad qualities can be formed.

4. Based on structure implies multiplicity (as illustrated by your plotted chord above), how many individual triads will be associated with each quality? (Describe in detail.)

    One triad of one quality, three triads of another quality, and three

    more triads of another quality can be formed.

5. Explain how you determined the answer to question 4, without writing out all of the triads.

    The distances between the dots around the circle (as labeled

    according to the circle of fifths) are 1, 3, and 3. These distances

    indicate exactly how many of each different triad quality can be

    formed.

One possible way to plot a seventh chord using the given notes on the circle diagram (arranged in the circle-of-fifths pattern) and answers to the questions in Exercise 3.5b

1.

A

D   E

G#   B

C#   F#

2. Based on cardinality equals variety, how many different seventh-chord qualities can be formed in the diatonic collection? ___4___

3. Explain how you determined this answer, without writing out all of the seventh chords.

Because there are four notes in a seventh chord, cardinality equals

variety indicates that there will be exactly four different interval

patterns formed by transpositions of the seventh chord. Therefore,

exactly four different seventh-chord qualities can be formed.

4. Based on structure implies multiplicity, how many individual seventh chords will be associated with each quality? (Describe in detail.)

One seventh chord of one quality, three seventh chords of another

quality, one seventh chord of another quality, and finally two

seventh chords of one more quality can be formed.

5. Explain how you determined this answer, without writing out all of the seventh chords.

The distances between the dots around the circle (as labeled

according to the circle of fifths) are 1, 3, 1, and 2. These distances

indicate exactly how many of each different triad quality can be

formed.

A Scale-Based View of Diatonic Chords

Exercise 3.6 provides an opportunity to test the assertions you made in the previous exercise by examining each individual chord quality in diatonic contexts. We will accomplish this test using a slightly different approach than we used in Chapter 2. Because we are dealing here with harmonic sonorities rather than melodic lines, we will proceed by writing chords on each note of the major scale rather than by transposing the initial formation, though essentially we still are just transposing the initial chord diatonically. Although the results, of course, are identical to those of the procedure employed in Chapter 2, taking a scalar approach may seem more musically intuitive in this case, particularly because theory students are often asked to produce such catalogues of diatonic triads and seventh chords by writing a chord on each note of a scale. Also, instead of labeling these chords with intervals between each note, as we did with the melodic lines in Chapter 2, we will identify these chords by chord root and quality. In so doing, we are simply generalizing the interval pattern of each chord. Finally, we will supply Roman numerals for each chord formed. In this way, you will have a direct opportunity to relate the work you will accomplish in this exercise to one of the primary methods of chord identification that you are likely to use in your further studies of music theory.

Roman Numerals for Triad Identification

Before embarking on this exercise, perhaps a brief introduction to or review of the use of Roman numerals for chord identification would be useful. First, the numerical value of a Roman numeral indicates the scale step on which a chord is built. For example, a chord built on the first note of the scale, or the tonic, is shown as I or i, and a chord built on the fourth note of the scale is labeled IV or iv. We will use uppercase Roman numerals to indicate major-quality chords and lowercase Roman numerals for minor quality chords. A diminished chord will be shown in lowercase with a superscript circle (o) to the right of the Roman numeral, and an augmented chord will be shown in uppercase with a superscript plus sign (+) to the right of the Roman numeral. Thus, a diminished triad built on the seventh note of the scale will be shown as vii°, and an augmented triad built on the third note of the scale will be shown as III$^+$.

Roman Numerals for Seventh Chord Identification

We will employ the same basic labeling scheme for seventh chords, except that we will indicate the quality of the seventh in each chord in the following way. A major seventh chord will be shown as an uppercase Roman numeral with a superscript "M7" to indicate the major seventh above the root, whereas a dominant seventh chord will be shown as an uppercase Roman numeral with a superscript "7" to imply a minor seventh above the root. Thus, a major seventh chord built on the first note of the scale will be shown as I^{M7}, whereas a dominant seventh chord built on the fifth note of the scale will be displayed as V^7. In like manner, a minor seventh chord, which has a minor seventh above the root, will be shown as a lowercase Roman numeral with a superscript "7," for example, ii^7.

Diminished seventh chords and half-diminished seventh chords will be treated separately. A diminished seventh chord will be shown as a lowercase Roman numeral with a superscript "o7," implying that both the triad quality and the quality of the interval of a seventh above the root are diminished. Finally, a half-diminished seventh chord will be shown as a lowercase Roman numeral with a superscript "ø7"—with the slash through the circle representing the idea of half diminished: a diminished triad with a minor seventh above the root. Thus, a diminished seventh chord built on the seventh note of the scale would be shown as vii^{o7}, and a half-diminished seventh chord built on the same note would be shown as $vii^{ø7}$.

There are many other schemes for labeling chords with Roman numerals. You may use whatever system you have learned; however, using some system that has distinct chord symbols for every different chord quality, such as the labeling system described above, will be essential to our discussion. Complete Exercise 3.6 using these identification guidelines (or some similar alternative system). At the conclusion of each part of this exercise is a broad question that gives you an opportunity to sum up your knowledge of cardinality equals variety and structure implies multiplicity by describing how these two properties apply to the triads and seventh chords that you have constructed. Rather than having specific, leading questions to guide you, you are left to make your own conclusions about what you are observing.

To test the assertions you made in Exercise 3.5a (for triads), examine the qualities of triads for the following indicated scales. First, write the triads formed on each note of the ascending A major scale in bass clef, using the appropriate key signature. Then, identify the root and quality of each chord, and indicate the appropriate Roman numeral for each. Repeat these steps with the other indicated scales and clefs.

A major:

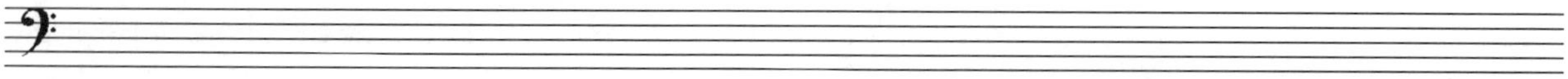

root/qual.:
Roman num.:

E♭ major:

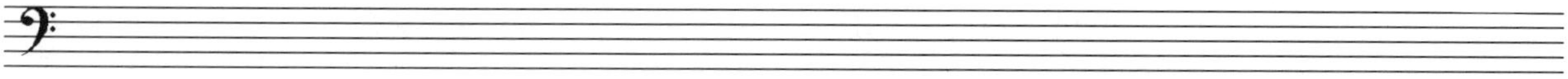

root/qual.:
Roman num.:

D major:

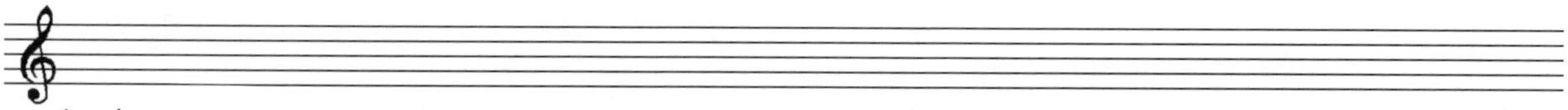

root/qual.:
Roman num.:

D♭ major:

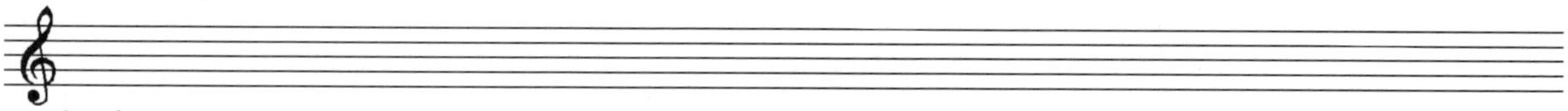

root/qual.:
Roman num.:

How does your work with the triads and scales in this exercise relate to cardinality equals variety and structure implies multiplicity? (Refer to Exercise 3.5a on p. 132.)

_____

_____

_____

_____

_____

_____

_____

_____

_____

EXERCISE 3.6b To test the assertions you made in Exercise 3.5b (for seventh chords), examine the qualities of seventh chords for the following indicated scales. First, write the seventh chords formed on each note of the ascending A major scale in bass clef, using the appropriate key signature. Then, identify the root and quality of each chord, and indicate the appropriate Roman numeral for each. Repeat these steps with the other indicated scales and clefs.

A major:

root/qual.:
Roman num.:

E♭ major:

root/qual.:
Roman num.:

D major:

root/qual.:
Roman num.:

D♭ major:

root/qual.:
Roman num.:

How does your work with the seventh chords and scales in this exercise relate to cardinality equals variety and structure implies multiplicity? (Refer to Exercise 3.5b on p. 133.)

Variety, Multiplicity, and Chord Identification

As shown in Solution 3.6a, only three diatonic triads can be formed—diminished, major, and minor. Thus, cardinality (three notes in a triad) equals variety (three triad qualities can be formed). Furthermore, there is one diminished triad (vii°), three major triads (I, IV, and V), and three minor triads (ii, iii, and vi)—as predicted in Exercise 3.5a by the distances between the dots associated with a triad (1–3–3). Therefore, structure (of a triad in relation to the circle of fifths) implies multiplicity (the number of each triad quality that can be formed). Consequently, the familiar pattern of Roman numerals for triads formed on the notes of the major scale (I, ii, iii, IV, V, vi, vii°) is directly connected to the properties we have been learning in this book. Cardinality equals variety indicates exactly how many triad qualities will occur, and structure implies multiplicity indicates how many chords of each quality will be formed. Play these triads on the piano, skipping freely among the various chords, so that you can aurally experience the results of this exercise.

As shown in Solution 3.6b, only four diatonic seventh chords can be formed—dominant, minor, half-diminished, and major. Again, cardinality (four notes in a seventh chord) equals variety (four seventh-chord qualities can be formed). There is one dominant seventh chord (V^7), three minor seventh chords (ii^7, iii^7, and vi^7), one half-diminished seventh chord ($vii^{ø7}$), and two major seventh chords (I^{M7}, IV^{M7})—as predicted in Exercise 3.5b by the distances between the dots associated with a seventh chord (1–3–1–2). Therefore, structure (of a seventh chord in relation to the circle of fifths) implies multiplicity (the number of each seventh-chord quality that can be formed). Here again, the familiar pattern of Roman numerals for seventh chords formed on the notes of the major scale (I^{M7}, ii^7, iii^7, IV^{M7}, V^7, vi^7, $vii^{ø7}$) is directly connected to the properties at hand. Cardinality equals variety indicates exactly how many seventh-chord qualities will occur, and structure implies multiplicity indicates how many chords of each quality will be formed. Play these seventh chords on the piano, skipping freely among the various chords, so that you can aurally experience the results of this exercise.

Roman Numerals and Other Scales

As you have observed in these exercises, the properties we have tested and the chord qualities resulting from building a chord on each note of the scale are consistent, regardless of which major scale is used. Furthermore, the same properties will hold true if the natural minor scale is used—or, in fact, if any of the seven modes are used with chords built on each note. However, because the harmonic minor scale is the primary context for building harmonies in a minor key, and because this form of the minor scale does not exhibit cardinality equals variety or structure implies multiplicity (as we discovered in Chapter 2), we will not explore the Roman numerals associated with minor scales here. Similarly, because chords built on the modes are sometimes not labeled with Roman numerals in theoretical discourse, and especially because further study in this manner would not reveal additional facts about the diatonic collection, we will not explore the structure and configuration of chords built on the modes either. If you wish, you can construct chords on each note of any mode to verify that the same chord qualities that

The triads formed on each note of the indicated scales in Exercise 3.6a and the relationships among these triads and the concepts of cardinality equals variety and structure implies multiplicity

A major:

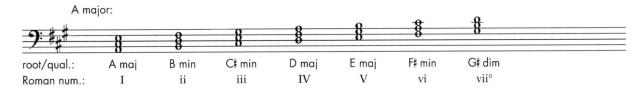

root/qual.:	A maj	B min	C♯ min	D maj	E maj	F♯ min	G♯ dim
Roman num.:	I	ii	iii	IV	V	vi	vii°

E♭ major:

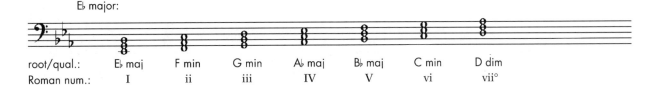

root/qual.:	E♭ maj	F min	G min	A♭ maj	B♭ maj	C min	D dim
Roman num.:	I	ii	iii	IV	V	vi	vii°

D major:

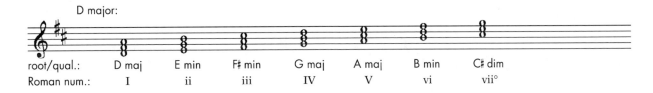

root/qual.:	D maj	E min	F♯ min	G maj	A maj	B min	C♯ dim
Roman num.:	I	ii	iii	IV	V	vi	vii°

D♭ major:

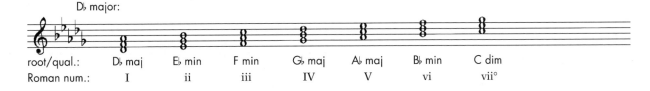

root/qual.:	D♭ maj	E♭ min	F min	G♭ maj	A♭ maj	B♭ min	C dim
Roman num.:	I	ii	iii	IV	V	vi	vii°

How does your work with the triads and scales in this exercise relate to cardinality equals variety and structure implies multiplicity?

According to cardinality equals variety, three triad qualities can be formed. The scales shown here each contain exactly three different triad qualities. According to structure implies multiplicity, the number of triads corresponding to each quality can be determined by the distances between notes around the circle of fifths (shown in Exercise 3.5a). The scales shown here each contain exactly one diminished triad, three major triads, and three minor triads, as indicated by the distances of 1, 3, and 3 between notes around the circle of fifths.

The seventh chords formed on each note of the indicated scales in Exercise 3.6b and the relationships among these seventh chords and the concepts of cardinality equals variety and structure implies multiplicity

A major:

root/qual.:	A maj7	B min7	C♯ min7	D maj7	E dom7	F♯ min7	C♯ half-dim7
Roman num.:	I^{M7}	ii^7	iii^7	IV^{M7}	V^7	vi^7	vii$^{∅7}$

E♭ major:

root/qual.:	E♭ maj7	F min7	G min7	A♭ maj7	B♭ dom7	C min7	D half-dim7
Roman num.:	I^{M7}	ii^7	iii^7	IV^{M7}	V^7	vi^7	vii$^{∅7}$

D major:

root/qual.:	D maj7	E min7	F♯ min7	G maj7	A dom7	B min7	C♯ half-dim7
Roman num.:	I^{M7}	ii^7	iii^7	IV^{M7}	V^7	vi^7	vii$^{∅7}$

D♭ major:

root/qual.:	D♭ maj7	E♭ min7	F min7	G♭ maj7	A♭ dom7	B♭ min7	C half-dim7
Roman num.:	I^{M7}	ii^7	iii^7	IV^{M7}	V^7	vi^7	vii$^{∅7}$

How does your work with the seventh chords and scales in this exercise relate to cardinality equals variety and structure implies multiplicity?

According to cardinality equals variety, four seventh-chord qualities can be formed. The scales shown here each contain exactly four different seventh-chord qualities. According to structure implies multiplicity, the number of seventh chords corresponding to each quality can be determined by the distances between notes around the circle of fifths (shown in Exercise 3.5b). The scales shown here each contain exactly one dominant seventh chord, three minor seventh chords, one half-diminished seventh chord, and two major seventh chords, as indicated by the distances of 1, 3, 1, and 2 between notes around the circle of fifths.

we observed in this exercise will be formed, as stipulated by the properties we have been discussing, except that the various chord qualities will appear on different scale steps than those we observed in the major scale.

Summary and Extensions

This chapter has focused on the primary harmonic building-blocks of diatonic music. We have observed the second-order maximal evenness of diatonic triads and seventh chords in relation to the diatonic collection, and we have verified that cardinality equals variety and structure implies multiplicity for triads and seventh chords drawn from the diatonic collection. In addition to offering an opportunity to explore these fundamental principles further, this chapter has also provided a thorough introduction to the structure and configuration of diatonic triads and seventh chords in general. Although we have worked with the same concepts and definitions as in Chapters 1 and 2, we have focused exclusively on triads and seventh chords in this chapter because these harmonic constructs are essential components of diatonic music theory.

Augmented Triads and Diminished Seventh Chords in Context

In contrast to diatonic triads and seventh chords, we again observed that the augmented triad and diminished seventh chord are both maximally even with respect to the chromatic universe but not with respect to the diatonic collection. It is ironic that common-practice composers seem to have saved the augmented triad for more chromatic contexts, whereas they began to adopt the diminished seventh chord into their otherwise primarily diatonic-based musical language much sooner. When you have the opportunity to explore this apparent disparity further in later theory courses—where the highly charged nature of the diminished seventh chords likely will be examined in more detail and in musical contexts—the background you have established here from working with these chords will be invaluable. For now, however, we must be content with the abstract observations that we have made concerning the relationship among all of these chords, based on the fact that they are all maximally even harmonic structures in one way or another. As in the other chapters in this book, we have not examined triads and seventh chords in their musical contexts. However, in general, having this opportunity to explore the abstract principles underlying triads and seventh chords may help you to solidify your conceptualization of the structure of these chords, before you embark on observations about the nature and behavior of these chords in musical contexts through your further studies in music theory.

Maximal Evenness and Physics

In this chapter, and especially in the first chapter, we have observed the importance of maximal evenness for diatonic musical structures. We have seen the nested structure of diatonicism, where triads and seventh chords are maximally even with respect to the diatonic collection, and where the diatonic collection is itself maximally even with respect to the chromatic universe. The musical import is clear—but what may be surprising is that scholars have found that the concept of maximal evenness also can be employed in statistical mechanics, a subfield of mathematical physics.[5]

Maximal Evenness and the Ising Model

The *Ising model* is a useful construct for modeling the behavior of electrons, though it may be applied in numerous other ways as well. Although it is not restricted to this usage alone, "the Ising model is a widely used standard model of statistical physics. Each year, about 800 papers are published that use the model to address problems in such diverse fields as neural networks, protein folding, biological membranes and social behavior."[6] The one-dimensional version of the model may be displayed in the form of a line extending in both directions to infinity, or in the form of a closed circle diagram.[7] In this model, electrons, placed around the circle, are spinning in one direction or the other (called *up-spins* and *down-spins*). In one version of the model, called *antiferromagnetic*, the up-spins and down-spins preferentially alternate, thus minimizing the overall energy of the system based on the configuration of up- and down-spins (called *configurational energy*). Because there may not be an equal number of up- and down-spins (for example, there may be more down-spins than up-spins in a given model), the electrons cannot simply line up with alternating spins. However, the optimum arrangement of up- and down-spins can be determined by applying the concept of maximal evenness from music theory scholarship to this model.[8] A maximally even distribution of up- and down-spins minimizes the configurational energy. Furthermore, with the application of an outside magnetic field, a standard procedure in working with Ising models that changes the spin behavior of the electrons substantially, the electron alignment that produces the minimum average energy is still maximally even.[9]

Although the history of the relationship between mathematics and music is long and varied, as outlined in the introduction to this book, for the most part this interaction has involved using mathematical constructs and ideas to help explain music. But in this case, the roles are reversed, and music theory is being used to help elucidate aspects of an important model in mathematical physics.

CONCLUSION

DO WE NOW HAVE ANY ANSWERS?

At this point it seems appropriate to reflect back on the questions posed at the beginning of this book. Why does the major scale seem to work so well? Why has diatonicism formed the backbone of Western music for so long—in both classical and popular music? And, perhaps most acutely, why are the black and white keys of the piano arranged in that way? By completing the exercises in this book, you have shed some light on these important questions, though we may never be able to answer them definitively and totally.

Review of Chapter 1

Chapter 1 introduced maximally even collections with respect to the twelve-note chromatic universe. The main goal of this chapter was to learn about the structure of the diatonic collection and to contemplate its special arrangement of whole and half steps. Another goal was to determine what other musical structures have a comparable spatial layout to the diatonic collection, or in other words, are maximally even. The chapter provided opportunities for you to explore an abstract representation of notes (a circle diagram with dots strategically placed around it to represent various musical structures) and to work with intervals within that abstract representation, rather than primarily on a staff or a keyboard. In addition to your work with maximally even structures, you also had a chance to consider intervallic relationships among the notes of the harmonic and melodic minor scales in relation to maximal evenness, helping you to observe idiosyncratic aspects of these two collections. Finally, you were introduced briefly to the deep scale property—another defining property of the diatonic collection, where every interval appears within the collection a unique number of times.

Review of Chapter 2

Chapter 2 introduced the properties of cardinality equals variety and structure implies multiplicity. In this chapter you learned that the diatonic, pentatonic, and whole tone collections can be produced by a generator and are well formed, whereas some other collections can be produced only by a bisector (a weaker surrogate). You worked with intervals among the notes of the diatonic collection, further familiarizing yourself with the nature of the pattern of whole and half steps intrinsic to this collection. In reference to the diatonic collection, you observed that the number of notes contained in any particular series drawn from this collection precisely indicates the number of distinct interval patterns that diatonic transpositions of that series will form (or cardinality equals variety). Furthermore, you observed that the number of times each interval pattern appears within a group of transposed series of notes can be determined by computing the distances between the notes of the original series, as placed around the (generated) circle of fifths (or structure implies multiplicity). Finally, you observed that the diatonic and pentatonic collections each have exactly two c distances for every d distance (or Myhill's property) and that this special intervallic structure yields the other two properties.

In this chapter you also transposed collections of notes diatonically, obtained practice in interval identification through exercises that had a collateral goal, and developed your own hypotheses about what you observed in your work. Furthermore, by considering the intervallic patterns formed by various transpositions of a pattern and by noting the number of different interval patterns that appear in each group, you had an opportunity to develop a broader understanding of diatonicism. By exploring these ideas in a self-directed way, you may have felt that you were able to grasp the theoretical ideas more firmly, and your resulting "ownership" of these ideas may facilitate further exploration of such generalized concepts in your future studies.

Review of Chapter 3

In Chapter 3 you applied the primary properties you learned in the previous two chapters to triads and seventh chords, perhaps the two most important and familiar harmonic sonorities to students of music theory. In this investigation you were asked to adapt the concept of maximal evenness to a diatonic context to determine if diatonic triads and seventh chords are maximally even with respect to the diatonic collection rather than to the chromatic universe (or second-order maximal evenness). You also had an opportunity to explore the configuration of triad and seventh-chord qualities with reference to the major scale, using Roman numerals to identify these chords. In so doing, you learned that cardinality equals variety and structure implies multiplicity for diatonic triads and seventh chords with respect to the diatonic collection. This nested configuration underlies one of the primary foundations upon which diatonic music is based, with its heavy reliance on triads and seventh chords for harmonic sonorities. Recognizing that the three primary properties that we have studied in this book apply to these chords is essential to our understanding of the fundamental importance of diatonic triads and seventh chords.

Toward Future Course Work

In addition to the immediate benefits of encountering these ideas at the beginning level of music theory, you are likely to find that these ideas resurface naturally in later theory courses. For example, understanding the maximally even internal structure of the diminished seventh chord is indispensable to the study of enharmonic modulation. Because you already have worked with the basic configuration of this chord and have noted its maximally even structure (and therefore, its inherent symmetry), you may be able to grasp the concept of enharmonic modulation, based on reinterpreting the diminished seventh chord in various keys, more easily and quickly when you encounter this idea in an advanced music theory course.

The material you have studied also begins to prepare you for future course work in twentieth-century music. By using an approach based on pitch-class set theory, the material in this book has provided a general introduction to some of the basic theoretical tools for the study of atonal music. Although this text has focused primarily on diatonically oriented structures, many of the basic procedures employed in these exercises—such as working with circle diagrams, identifying intervals by counting half steps, and reaching generalizations based on observation—are directly applicable to the study of pitch-class set theory in association with atonal music. In addition, this book has shown how the concepts explored relate to standard terminology from pitch-class set theory whenever possible.

More Questions

Do we now have any answers to our original questions? Is there something special about the diatonic collection? And do you now have any ideas about why the diatonic collection—and its characteristic representation in the orientation of black and white keys on the piano keyboard—is arranged in the way that it is? Clearly, the special status of the diatonic collection is due in part to the fact that it is maximally even and that cardinality equals variety and structure implies multiplicity for any group of notes found within this collection. Further, the diatonic collection is a deep scale that is generated, well formed, and has Myhill's property. By working through this text you have observed some fundamental characteristics of this significant collection, though certainly much room remains for investigation of additional special properties. As you continue to contemplate the nature of the diatonic collection, you may form your own ideas about what makes this collection so special. Or, more importantly, you may begin to formulate additional questions about the diatonic collection and other musical constructs—questions that you may wish to try to solve yourself.

Do you have any questions?

FOR FURTHER STUDY

Students who wish to delve deeper into the ideas that have been presented in this text may wish to consult the original sources of this pedagogical material. Introductory students, for whom this book is primarily intended, should be cautioned, however. Much of the material presented in the articles noted here is couched in the formal language of mathematical discourse. Although this textbook has introduced you to the ideas contained in some of these articles, it has not attempted to prepare you for the very technical manner of presentation. Nevertheless, currently there is no intermediate-level treatment of this material, and students who wish to enrich their understanding of diatonic set theory would need to consult the original scholarship, as outlined below.

JOHN CLOUGH

As mentioned in the Preface, the material in this textbook draws upon the work of John Clough and his various collaborators. Clough continues to be a leading voice in the field, and his list of articles that are relevant to the study of diatonic theory is extensive. Clough's work ranges from "Aspects of Diatonic Sets," where he laid out some initial observations following a mathematically based approach to diatonic structures, to his recent collaborative work, "Scales, Sets, and Interval Cycles: A Taxonomy," which draws together numerous ideas and consolidates a wealth of information from the field.

Suggested Reading

Clough, John. "Aspects of Diatonic Sets." *Journal of Music Theory* 23 (1979): 45–61.

———. "Diatonic Interval Sets and Transformational Structures." *Perspectives of New Music* 18, no. 1–2 (1979–80): 461–82.

Clough, John, and Gerald Myerson. "Variety and Multiplicity in Diatonic Systems." *Journal of Music Theory* 29 (1985): 249–70.

Clough, John, and Jack Douthett. "Maximally Even Sets." *Journal of Music Theory* 35 (1991): 93–173.

Clough, John, Jack Douthett, N. Ramanathan, and Lewis Rowell. "Early Indian Heptatonic Scales and Recent Diatonic Theory." *Music Theory Spectrum* 15 (1993): 36–58.

Clough, John. "Diatonic Interval Cycles and Hierarchical Structure." *Perspectives of New Music* 32, no. 1 (1994): 228–53.

Clough, John, John Cuciurean, and Jack Douthett. "Hyperscales and the Generalized Tetrachord." *Journal of Music Theory* 41 (1997): 67–100.

Clough, John, Nora Engebretsen, and Jonathan Kochavi. "Scales, Sets, and Interval Cycles: A Taxonomy." *Music Theory Spectrum* 21 (1999): 74–104.

$\mathcal{M}$AXIMAL EVENNESS

Those who would like to discover the origins of the term maximal evenness firsthand should consult "Maximally Even Sets," written by Clough in collaboration with mathematician Jack Douthett. The material in Chapter 1 and some of the material in Chapter 3 derive chiefly from this award-winning article. (This groundbreaking article was recognized by the music theory community with a Society for Music Theory Publication Award in 1995.) The article goes far beyond the presentation of the topic in this textbook, both in scope and in its focus on mathematical proofs.

Other scholars have followed up on Clough and Douthett's original research. For example, Steven Block and Jack Douthett provided a geometrically based definition of maximal evenness. Interested students may wish to compare these two vastly different approaches to the same musical concept. And as mentioned in Chapter 3, Jack Douthett, Richard Krantz, and Steven Doty have expanded the reach of maximal evenness to forge relationships with another discipline. Their work has shown important correlations between properties of diatonic music and models of electron behavior in the physical sciences.

Suggested Reading

Clough, John, and Jack Douthett. "Maximally Even Sets." *Journal of Music Theory* 35 (1991): 93–173.

Block, Steven, and Jack Douthett. "Vector Products and Intervallic Weighting." *Journal of Music Theory* 38 (1994): 21–41.

Krantz, Richard, Jack Douthett, and John Clough. "Maximally Even Sets: A Discovery in Mathematical Music Theory is Found to Apply in Physics." In *Bridges: Mathematical Connections in Art, Music, and Science*. Conference Proceedings, ed. Reza Sarhangi, 193–200. Winfield, Kansas: Central Plain Book Manufacturing, 2000.

Douthett, Jack, and Richard Krantz. "Energy Extremes and Spin Configurations for the One-Dimensional Antiferromagnetic Ising Model with Arbitrary-Range Interaction." *Journal of Mathematical Physics* 37 (1996): 3334–53.

Krantz, Richard J., Jack Douthett, and Steven D. Doty. "Maximally Even Sets and the Devil's-Staircase Phase Diagram for the One-Dimensional Ising Antiferromagnet with Arbitrary-Range Interaction." *Journal of Mathematical Physics* 39 (1998): 4675–82.

VARIETY AND MULTIPLICITY

For a glimpse of the original conception of the other two primary principles explored in this textbook, cardinality equals variety and structure implies multiplicity, consult the pioneering article, "Variety and Multiplicity in Diatonic Systems," by John Clough and Gerald Myerson. Chapter 2 and parts of Chapter 3 of this textbook draw primarily from this article, which presented these concepts for the first time.

Subsequently, Eytan Agmon has explored these ideas independently, as well as some other fundamental principles of diatonicism, by means of an entirely new approach and using alternative terms. Later he attempted to find some common ground among divergent terminologies and approaches to diatonic theory. Although his work differs significantly from the terminology adopted in this textbook, his articles present opportunities to view some of the same basic concepts in a new light.

Suggested Reading

Clough, John, and Gerald Myerson. "Variety and Multiplicity in Diatonic Systems." *Journal of Music Theory* 29 (1985): 249–70.
Agmon, Eytan. "A Mathematical Model of the Diatonic System." *Journal of Music Theory* 33 (1989): 1–25.
———. "Coherent Tone-Systems: A Study in the Theory of Diatonicism." *Journal of Music Theory* 40 (1996): 39–59.

OTHER DEVELOPMENTS

In addition to those scholars who have developed theories on maximal evenness, cardinality equals variety, and structure implies multiplicity—the primary concerns of this textbook—others have contributed significantly to the development of the field. Scholars have been active in this field, as scholarship on diatonic set theory grows and the number of theorists interested in this area expands rapidly. For example, Norman Carey and David Clampitt discovered some new properties of diatonicism—including generated and well-formed scales, as discussed in Chapter 2 of this textbook. Although the scholarship in the following list varies widely in approach and difficulty, all of these articles reveal the same kind of curiosity about the diatonic collection that has been fostered throughout this textbook.

Suggested Reading

Carey, Norman, and David Clampitt. "Aspects of Well-Formed Scales." *Music Theory Spectrum* 11 (1989): 187–206.
Carey, Norman, and David Clampitt. "Self-Similar Pitch Structures, Their Duals, and Rhythmic Analogues." *Perspectives of New Music* 34, no. 2 (1996): 62–87.
Browne, Richmond. "Tonal Implications of the Diatonic Set." *In Theory Only* 5, no. 6–7 (1981): 3–21.

Rahn, Jay. "Coordination of Interval Sizes in Seven-Tone Collections." *Journal of Music Theory* 35 (1991): 33–60.

Balzano, Gerald J. "The Group-Theoretic Description of 12-fold and Microtonal Pitch Systems." *Computer Music Journal* 4 (1980): 66–84.

———. "The Pitch Set as a Level of Description for Studying Musical Pitch Perception." In *Music, Mind, and Brain: The Neuropsychology of Music*, ed. Manfred Clynes, 321–51. New York: Plenum Press, 1982.

Brinkman, Alexander R. "A Binomial Representation of Pitch for Computer Processing of Musical Data." *Music Theory Spectrum* 8 (1986): 44–57.

SOME PRECURSORS

Students who wish to trace the historical development of diatonic set theory might begin with Milton Babbitt, an important American composer and theorist. He appears to have been the first to suggest that diatonic music might be explored by means of mathematically oriented procedures that previously had been limited to post-tonal music. Although his primary focus was electronic music, his article has been enormously influential in this field. Later, Carlton Gamer explored some fundamental aspects of the structure and nature of the diatonic collection—in particular, the notion of deep scales, discussed at the end of Chapter 1 in this textbook. Other authors, including Eric Regener and Jay Rahn, followed suit, eventually leading to an explosion of interest in diatonic set theory that was fueled to a great extent by the work of John Clough.

Suggested Reading

Babbitt, Milton. "Twelve-Tone Rhythmic Structure and the Electronic Medium." *Perspectives of New Music* 1, no. 1 (1962): 49–79.

Gamer, Carlton. "Deep Scales and Difference Sets in Equal-Tempered Systems." *American Society of University Composers: Proceedings of the Second Annual Conference* (1967): 113–22.

———. "Some Combinational Resources of Equal-Tempered Systems." *Journal of Music Theory* 11 (1967): 32–59.

Regener, Eric. "On Allen Forte's Theory of Chords." *Perspectives of New Music* 13, no. 1 (1974): 191–212.

Rahn, Jay. "Some Recurrent Features of Scales." *In Theory Only* 2, no. 11–12 (1977): 43–52.

DIATONIC THEORY AND HISTORICAL STUDIES

Although diatonic set theory has had a very short history, many of the ideas that have been developed are also relevant to music and musical discourse of earlier times. A number of scholars have displayed a keen interest in observing examples of early music and early music treatises in the context of diatonic set theory. Jay Rahn explored modal music of the fourteenth and early fifteenth centuries through a mathematically oriented approach, and

Robert Gauldin associated aspects of diatonic set theory with ancient systems of tone relations. Norman Carey and David Clampitt likewise expanded the historical scope of diatonic set theory by attempting to understand the work of early medieval theorists in light of recent developments in the study of diatonicism.

Suggested Reading

Rahn, Jay. "Constructs for Modality, Ca. 1300–1550." *Canadian Association of University Schools of Music Journal* 8, no. 2 (1978): 5–39.

Gauldin, Robert. "The Cycle-7 Complex: Relations of Diatonic Set Theory to the Evolution of Ancient Tonal Systems." *Music Theory Spectrum* 5 (1983): 39–55.

Carey, Norman, and David Clampitt. "Regions: A Theory of Tonal Spaces in Early Medieval Treatises." *Journal of Music Theory* 40 (1996): 113–47.

DIATONIC THEORY AND NON-WESTERN MUSIC

Scholars have begun to examine scale systems of non-Western cultures using tools and approaches derived from diatonic set theory. For example, John Clough and several others representing diverse backgrounds have collaborated to show that some of the principles observed in diatonic Western scales are also present in scales of ancient and medieval India. In addition, though their article focuses primarily on Western scales, Norman Carey and David Clampitt asserted that concepts they developed in connection with the twelve-note chromatic universe apply equally well to the seventeen-tone Arabic and fifty-three–tone Chinese scalar systems. Clearly, much remains to be known about the relationship between the theoretical concepts that have been presented in this textbook and the scalar constructs of non-Western cultures. Yet, the fact that at least some of these theoretical constructs seem to cross cultural borders strongly supports the idea that music, at least on some level, is universal.

Suggested Reading

Clough, John, Jack Douthett, N. Ramanathan, and Lewis Rowell. "Early Indian Heptatonic Scales and Recent Diatonic Theory." *Music Theory Spectrum* 15 (1993): 36–58.

Clough, John, John Cuciurean, and Jack Douthett. "Hyperscales and the Generalized Tetrachord." *Journal of Music Theory* 41 (1997): 67–100.

Carey, Norman, and David Clampitt. "Aspects of Well-Formed Scales." *Music Theory Spectrum* 11 (1989): 187–206.

ANALYSIS

Finally, a largely untapped area of inquiry, in regard to diatonic set theory, is musical analysis. The theoretical constructs presented in this textbook, and in most of the research cited previously in this section, deal primarily with abstract considerations. Few scholars have attempted to show how these theoretical ideas might be applied to the analysis of musical literature, and this textbook has been no exception in this regard as it has dealt with foundations rather than applications of diatonic theory. On the contrary, Matthew Santa, in an insightful article, studied diatonic, post-tonal music by twentieth-century composers—including Igor Stravinsky, Samuel Barber, and Sergey Prokofiev—through an approach based on diatonic set theory. John Clough also provided some brief analytical remarks on works by Mozart and Beethoven; however, his article, as with most of his writing, is chiefly theoretical rather than analytical. As the field becomes more widely known, beginning to a certain extent with the students who first encounter some of the basic tenets of the theory in this textbook, it seems certain that more analytical applications of diatonic set theory will begin to appear.

Suggested Reading

Santa, Matthew. "Analysing Post-Tonal Diatonic Music: A Modulo 7 Perspective." *Music Analysis* 19 (2000): 167–201.

Clough, John. "Aspects of Diatonic Sets." *Journal of Music Theory* 23 (1979): 45–61.

COMING BACK TO DIATONIC SET THEORY

In addition to its primary objective, the study of music fundamentals in relation to specific aspects of diatonic set theory, this textbook also may have sparked a broader interest in exploring some of the highly specialized literature cited earlier or in finding creative new ways to contemplate music. Although the scholars listed in this section of the textbook are operating at a very high and intense level in many respects, developing a curiosity about their work may pay dividends later as you continue your studies in music theory. Delving into the history of ideas outlined in the suggested readings can provide keen insights that otherwise might be missed. Likewise, coming back to this textbook as you gain more knowledge and sophistication in music theory may provide you with a new appreciation of the ideas that you have developed by working through this material. In time, you may find yourself drawn anew to these concepts, and perhaps you will look beyond the abstract approach provided in this textbook to discover new theoretical or analytical applications for the foundations of diatonic theory that you have only begun to experience.

NOTES

Introduction

1. Edward Rothstein, *Emblems of Mind: The Inner Life of Music and Mathematics* (New York: Times Books, 1995). Rothstein provides a delightful, illuminating, and very accessible introduction to the interrelationship between mathematics and music, and in this regard he mentions relevant aspects of number theory and group theory, the subfields of mathematics upon which this textbook mainly relies.

2. Recent books that explore the connections between mathematics and music in general include: Robin Maconie, *The Science of Music* (New York: Oxford University Press, 1997); and Charles Madden, *Fractals in Music: Introductory Mathematics for Musical Analysis* (Salt Lake City: High Art Press, 1999).

3. Richard L. Crocker, "Pythagorean Mathematics and Music (Parts I & II)," *Journal of Aesthetics and Art Criticism* 22, no. 2–3 (1963–64): 334; Reprint: *Studies of Medieval Music Theory and the Early Sequence* (Brookfield, Vermont: Variorum, Ashgate Publishing Company, 1997).

4. André Barbera, "Pythagoras," in *The New Grove Dictionary of Music and Musicians,* 2d ed., ed. Stanley Sadie (London: Macmillan Publishers, 2001); and Mark Lindley, "Pythagorean Intonation," in *New Grove Dictionary.*

5. For a historical overview of the various temperaments that have been used with some regularity, see Mark Lindley, "Temperaments," in *New Grove Dictionary.*

6. Likewise, Gerald Balzano makes the specific point that many important properties of the diatonic and chromatic collections are independent of any concerns about ratios (Gerald J. Balzano, "The Group-Theoretic Description of 12-fold and Microtonal Pitch Systems," *Computer Music Journal* 4 [1980]: 66–84; Gerald J. Balzano, "The Pitch Set as a Level of Description for Studying Musical Pitch Perception," in *Music, Mind, and Brain: The Neuropsychology of Music,* ed. Manfred Clynes [New York: Plenum Press, 1982], 321–51). On the other hand, Norman Carey and David Clampitt attempt to reconcile Pythagorean concepts of octave, fifth, and other intervals with some of their theories about diatonicism. But in so doing, they too demonstrate the independence of their formal theories from any particular tuning system employed (Carey and Clampitt, "Aspects of Well-Formed Scales," *Music Theory Spectrum* 11 [1989]: 194–200).

7. For example, see Carlton Gamer, "Deep Scales and Difference Sets in Equal-Tempered Systems," *American Society of University Composers: Proceedings of the Second Annual Conference* (1967): 113–22; Carlton Gamer, "Some Combinational Resources of Equal-Tempered Systems," *Journal of Music Theory* 11 (1967): 32–59; Richard J. Krantz and Jack Douthett, "A Measure of the Reasonableness of Equal-Tempered Musical Scales," *Journal of the Acoustical Society of America* 95 (1994): 3642–50; and John Clough, Nora Engebretsen, and Jonathan Kochavi, "Scales, Sets, and Interval Cycles: A Taxonomy," *Music Theory Spectrum* 21 (1999): 74–104.

8. This survey is based primarily on material from Claude V. Palisca and Ian D. Bent, "Theory, Theorists," in *New Grove Dictionary*. A more focused survey and commentary on the history of the relationship between mathematics and music appears in David Loeb, "Mathematical Aspects of Music," in *The Music Forum*, vol. 2., ed. William J. Mitchell and Felix Salzer (New York: Columbia University Press, 1970), 110–29.

9. C. André Barbera, "Arithmetic and Geometric Divisions of the Tetrachord," *Journal of Music Theory* 21 (1977): 294–323.

10. Sigalia Dostrovsky, Murry Campbell, James F. Bell, and C. Truesdell, "Physics of Music," in *New Grove Dictionary*.

11. Martin Scherzinger, "The Changing Roles of Acoustics and Mathematics in Nineteenth-Century Music Theory and Their Relation to the Aesthetics of Autonomy," *South African Journal of Musicology* 18 (1998): 21.

12. Dostrovsky, Campbell, Bell, and Truesdell, "Physics of Music."

13. An interesting account of scientific aspects of sound and music, directed expressly toward musicians rather than scientists, appears in Ian Johnston, *Measured Tones: The Interplay of Physics and Music* (New York: Adam Hilger, 1989). Another excellent book with similar aims, though currently out of print, is Siegmund Levarie and Ernst Levy, *Tone: A Study in Musical Acoustics*, 2d ed. (Kent, Ohio: The Kent State University Press, 1980). A more traditional scientific approach is taken by Johan Sundberg, *The Science of Musical Sounds* (San Diego: Academic Press, 1991). A fascinating interdisciplinary presentation—involving physics, acoustics, psychophysics, and neuropsychology—appears in Juan G. Roederer, *The Physics and Psychophysics of Music: An Introduction*, 3d ed. (New York: Springer, 1995). In *Tuning, Timbre, Spectrum, Scale* (New York: Springer, 1998), William A. Sethares explores ways to interrelate alternative scales and tunings with the study of acoustics. A good reference work for the study of acoustics, especially as associated with the various families of instruments, appears in Malcolm J. Crocker, ed., "Part XVI: Music and Musical Acoustics," in *Encyclopedia of Acoustics*, v. 4 (New York: John Wiley & Sons, 1997), 1615–95. This encyclopedia, in general, provides an excellent overview of the wide range of subtopics within the field of acoustics, ranging from basic theories of sound waves to architectural design.

14. Mathematics has been employed extensively in other areas besides pitch, but since this book is devoted solely to pitch—rather than rhythm, timbre, proportion, and other musical aspects—a survey of the relationship between these other musical components and mathematics will not be undertaken here. However, the interested reader might begin to explore some of these areas in Jonathan D. Kramer, *The Time of Music: New Meanings, New Temporalities, New Listening Strategies* (New York: Schirmer Books, 1988).

Chapter 1

1. A June 2000 issue of *Mathematics Magazine* challenged readers to prove that such a rounding procedure will always produce the result shown in Figure 1.1c (or one of its rotations shown in Solution 1.1c). The proposed proof uses white points and black points on a circle instead of lines, and employs an algebraic formula to produce the resulting figure, but the procedure is effectively the same (John Clough, Jack Douthett, and Roger Entringer, "Problem," *Mathematics Magazine* 73, no. 3 [June 2000]: 240).

2. John Clough has used a version of this analogy in a number of his lectures and presentations.

3. Eytan Agmon, "Coherent Tone-Systems: A Study in the Theory of Diatonicism," *Journal of Music Theory* 40 (1996): 39–59.

4. In the diatonic set theory literature, these distances often are referred to as *specific* and *generic intervals*, respectively.

5. In discussing the "largest" possible c or d distances, I am referring to clockwise distances between dots considered as ordered pairs.

6. Steven Block and Jack Douthett, "Vector Products and Intervallic Weighting," *Journal of Music Theory* 38 (1994): 35.

7. However, Block and Douthett identify this scale as the fourth-most maximally even seven-note collection, ranking just below the collection that contains the whole-tone scale plus one additional note (Ibid.).

8. The complement of the augmented triad—a nine-note maximally even scale, which is not discussed in this text—is *mode 3* of twentieth-century French composer Olivier Messiaen's *modes of limited transposition*.

9. Inversionally related pairs of intervals are called *interval classes,* in terms of pitch-class set theory, and are represented by the smaller interval of each pair.

10. In terms of pitch-class set theory, this table essentially constitutes an *interval-class vector*.

11. Gamer, "Deep Scales and Difference Sets"; and "Some Combinational Resources." Gamer attributed his contributions on deep scales to an unpublished paper by Terry Winograd ("An Analysis of the Properties of 'Deep Scales' in a T-Tone System," unpublished, n.d.). Later, Richmond Browne explored the significance of this special property in the structure of the diatonic collection (Browne, "Tonal Implications of the Diatonic Set," *In Theory Only* 5, no. 6–7 [1981]: 6–10).

12. This phenomenon is called the *common-tone theorem*.

13. Although modulations within musical compositions typically do not involve a literal change of key signature, new keys are implied by the chromatic notes that appear, and these new keys may be considered to be represented by the implied key signatures. The examples that follow are major keys, but comparisons between pairs of minor keys (using notes of the natural minor scale to determine common tones) work in the same way.

Chapter 2

1. Jay Rahn identifies three independent ways of approaching intervals in seven-note collections: (1) by half steps, (2) by scale steps, or (3) by connecting both half steps and scale steps (Jay Rahn, "Coordination of Interval Sizes in Seven-Tone Collections," *Journal of Music Theory* 35 [1991]: 34). The following exercises involve all three of these approaches.

2. Proving cardinality equals variety with respect to this or any other collection could be accomplished by exhaustively checking the interval patterns formed by every possible series of notes, or it could be accomplished mathematically. We will not attempt to prove cardinality equals variety here, but we will consider the examples solved as a sufficient demonstration of the property.

3. Eric Regener seems to have been the first to propose counting diatonic distances in terms of fifths, though the circle-of-fifths pattern itself has been well-known for centuries (Eric Regener, "On Allen Forte's Theory of Chords," *Perspectives of New Music* 13, no. 1 [1974]: 199–201). His brief section on diatonic chords in this article has been enormously influential in the development of the field of diatonic set theory.

4. Eytan Agmon uses the term *cyclic system* (Agmon, "Coherent Tone-Systems").

5. Gamer, "Some Combinational Resources," 41.

6. Carey and Clampitt, "Aspects of Well-Formed Scales."

7. Similarly, the same diatonic collection generated by a circle of perfect fourths would begin with B and work through the same series of notes in reverse order (c distance of 5, d distance of 3).

8. The same could be said of the circle of fourths, where a single fourth is augmented (c distance of 6, g + 1 or 5 + 1), and all of the other fourths are perfect (c distance of 5).

9. By way of review, if GCD of (c, d) = 1, then c and d are coprime. In this case GCD (12, 5) = 1.

10. The term *bisector* was introduced in Jay Rahn, "Some Recurrent Features of Scales," *In Theory Only* 2, no. 11–12 (1977): 43–52.

11. In his original conception of bisectors, Rahn applied the term to collections that are equally spaced, but I am adapting the term to apply to generic scale steps (or d distances) (Ibid., 45). Clough has made a similar adaptation (Clough, "Diatonic Interval Cycles and Hierarchical Structure" *Perspectives of New Music* 32, no. 1 [1994]: 235).

12. Rahn uses the term *aliquant bisector* for bisectors that can be used to produce every note of a collection. Only collections where the bisector (b) and the number of notes (d) are coprime will work in this way—or GCD (b, d) = 1 (Rahn, "Some Recurrent Features of Scales," 46). Thus, bisectors can be used to produce the diatonic, harmonic minor, and ascending melodic minor collections—GCD (4, 7) = 1 or GCD (3, 7) = 1.

13. Ibid., 45.

14. Scales with equal intervals, such as the whole-tone scale, are called *degenerate well-formed* scales because the generator and the interval required to complete the circle by returning to the initial note are equivalent (Carey and Clampitt, "Aspects of Well-Formed Scales," 200; and Clough, Engebretsen, and Kochavi, "Scales, Sets, and Interval Cycles," 79).

15. This property is named after John Myhill, a mathematician and associate of Clough and Myerson.

16. Clough, Engebretsen, and Kochavi, "Scales, Sets, and Interval Cycles," 78–84.

17. The interval patterns corresponding to these figures—2–7–3 for Figure 2.6a, 2–6–4 for Figure 2.6b, and 1–7–4 for Figure 2.6c—can be determined by writing out these series of notes on staff paper and identifying the intervals between notes, or the interval patterns can be calculated using the provided circles by means of modulo 12 arithmetic. To use modulo 12 arithmetic with these circles, sum the intervals of 7 (and the one 6 where applicable) that appear between each pair of dots, and divide each result by 12. The *remainders* correspond to the interval pattern for each circle. This procedure is left for you to explore on your own if you wish. Modular arithmetic is an important tool in diatonic set theory but remains primarily in the background in this textbook.

Chapter 3

1. This use of the term *inversion* is different from the one commonly associated with diatonic triads and seventh chords and is more closely aligned with the use of the term in conjunction with intervals that was employed earlier in this text. In traditional tonal theory, the inversion of chords involves revolving the notes by moving the bottom note in the chord so that it appears above the previous top note of the chord. For example, C–E–G becomes E–G–C. However, the application of the term inversion here involves changing the perspective of the notes relative to each other (or flipping). Thus, the bottom note in a chord maps into the top note, and the top note maps into the bottom note, and so forth until all notes have changed positions relative to each other.

2. It is conventional in set theory to invert around C—drawing the dashed line through C, rather than vertically through the diagram as in this demonstration. The line is drawn through E here because the principle is essentially the same, and inversion around a vertical line may be easier to see. However, following the

conventional procedure (inverting around C) would change only the interval of transposition (or rotation) to five half steps ascending (clockwise).

3. Inversionally related and transpositionally related chords, in terms of pitch-class set theory, are considered together as members of a single family, called a *set class*.

4. Although most musical structures that have identical interval contents are either transpositionally or inversionally related (or both), some pairs of musical structures with identical interval contents are not related in either of these ways. Structures that have identical interval contents but that are not related by transposition or inversion are called *Z-related*. For example, the collections C–C♯–E–F♯ and C–C♯–E♭–G are Z-related because they have identical interval contents (one instance of each interval from one to six half-steps), but these collections are not related by transposition or inversion. On the other hand, major/minor triads and dominant/half-diminished seventh chords, though they have identical interval contents, are *not* Z-related because they are *inversionally related* (a stronger affiliation). Musical structures that are Z-related likely will arise in your later studies of music theory in connection with pitch-class set theory and atonal music.

5. I would like to thank my colleague, James Conklin, Associate Professor of Mathematics at Ithaca College, who helped guide me through the relevant literature.

6. Conley Stutz and Beverly Williams, "Ernst Ising," Obituary, *Physics Today* 52, no. 3 (March 1999): 106.

7. Using a circle instead of a line, called *invoking periodic boundary conditions*, introduces a negligible degree of error into calculations for large numbers of sites around the circle in statistical applications of this model.

8. Jack Douthett and Richard Krantz, "Energy Extremes and Spin Configurations for the One-Dimensional Antiferromagnetic Ising Model with Arbitrary-Range Interaction," *Journal of Mathematical Physics* 37 (1996): 3334–53. Also, for a summary of applications of maximal evenness in physics and music theory, see Richard Krantz, Jack Douthett, and John Clough, "Maximally Even Sets: A Discovery in Mathematical Music Theory is Found to Apply in Physics," in *Bridges: Mathematical Connections in Art, Music, and Science*, Conference Proceedings 2000, ed. Reza Sarhangi (Winfield, Kansas: Central Plain Book Manufacturing, 2000), 193–200.

9. Richard J. Krantz, Jack Douthett, and Steven D. Doty, "Maximally Even Sets and the Devil's-Staircase Phase Diagram for the One-Dimensional Ising Antiferromagnet with Arbitrary-Range Interaction," *Journal of Mathematical Physics* 39 (1998): 4675–82.

Agmon, Eytan. "A Mathematical Model of the Diatonic System." *Journal of Music Theory* 33 (1989): 1–25.

_____. "Coherent Tone-Systems: A Study in the Theory of Diatonicism." *Journal of Music Theory* 40 (1996): 39–59.

Babbitt, Milton. "Twelve-Tone Rhythmic Structure and the Electronic Medium." *Perspectives of New Music* 1, no. 1 (1962): 49–79.

Balzano, Gerald J. "The Group-Theoretic Description of 12-fold and Microtonal Pitch Systems." *Computer Music Journal* 4 (1980): 66–84.

_____. "The Pitch Set as a Level of Description for Studying Musical Pitch Perception." In *Music, Mind, and Brain: The Neuropsychology of Music*, ed. Manfred Clynes, 321–51. New York: Plenum Press, 1982.

Barbera, C. André. "Arithmetic and Geometric Divisions of the Tetrachord." *Journal of Music Theory* 21 (1977): 294–323.

Barbera, André. "Pythagoras." In *The New Grove Dictionary of Music and Musicians*, 2d ed., ed. Stanley Sadie. London: Macmillan Publishers, 2001.

Block, Steven, and Jack Douthett. "Vector Products and Intervallic Weighting." *Journal of Music Theory* 38 (1994): 21–41.

Brinkman, Alexander R. "A Binomial Representation of Pitch for Computer Processing of Musical Data." *Music Theory Spectrum* 8 (1986): 44–57.

Browne, Richmond. "Tonal Implications of the Diatonic Set." *In Theory Only* 5, no. 6–7 (1981): 3–21.

Carey, Norman, and David Clampitt. "Aspects of Well-Formed Scales." *Music Theory Spectrum* 11 (1989): 187–206.

Carey, Norman, and David Clampitt. "Regions: A Theory of Tonal Spaces in Early Medieval Treatises." *Journal of Music Theory* 40 (1996): 113–47.

Carey, Norman, and David Clampitt. "Self-Similar Pitch Structures, Their Duals, and Rhythmic Analogues." *Perspectives of New Music* 34, no. 2 (1996): 62–87.

Clough, John. "Aspects of Diatonic Sets." *Journal of Music Theory* 23 (1979): 45–61.

_____. "Diatonic Interval Cycles and Hierarchical Structure." *Perspectives of New Music* 32, no. 1 (1994): 228–53.

_____. "Diatonic Interval Sets and Transformational Structures." *Perspectives of New Music* 18, no. 1–2 (1979–80): 461–82.

Clough, John, John Cuciurean, and Jack Douthett. "Hyperscales and the Generalized Tetrachord." *Journal of Music Theory* 41 (1997): 67–100.

Clough, John, and Jack Douthett. "Maximally Even Sets." *Journal of Music Theory* 35 (1991): 93–173.

Clough, John, Jack Douthett, and Roger Entringer. "Problem." *Mathematics Magazine* 73, no. 3 (June 2000): 240.

Clough, John, Jack Douthett, N. Ramanathan, and Lewis Rowell. "Early Indian Heptatonic Scales and Recent Diatonic Theory." *Music Theory Spectrum* 15 (1993): 36–58.

Clough, John, Nora Engebretsen, and Jonathan Kochavi. "Scales, Sets, and Interval Cycles: A Taxonomy." *Music Theory Spectrum* 21 (1999): 74–104.

Clough, John, and Gerald Myerson. "Variety and Multiplicity in Diatonic Systems." *Journal of Music Theory* 29 (1985): 249–70.

Cohn, Richard. "Music Theory's New Pedagogability." *Music Theory Online* 4.2 (1998).

Crocker, Malcolm J., ed. "Part XVI: Music and Musical Acoustics." In *Encyclopedia of Acoustics*, v. 4, 1615–95. New York: John Wiley & Sons, 1997.

Crocker, Richard L. "Pythagorean Mathematics and Music (Parts I & II)." *Journal of Aesthetics and Art Criticism* 22, no. 2–3 (1963–64): 189–98, 325–35. Reprint: *Studies of Medieval Music Theory and the Early Sequence*. Brookfield, Vermont: Variorum, Ashgate Publishing Company, 1997.

Dostrovsky, Sigalia, Murry Campbell, James F. Bell, and C. Truesdell. "Physics of Music." In *The New Grove Dictionary of Music and Musicians*, 2d ed., ed. Stanley Sadie. London: Macmillan Publishers, 2001.

Douthett, Jack, and Richard Krantz. "Energy Extremes and Spin Configurations for the One-Dimensional Antiferromagnetic Ising Model with Arbitrary-Range Interaction." *Journal of Mathematical Physics* 37 (1996): 3334–53.

Gamer, Carlton. "Deep Scales and Difference Sets in Equal-Tempered Systems." *American Society of University Composers: Proceedings of the Second Annual Conference* (1967): 113–22.

_____. "Some Combinational Resources of Equal-Tempered Systems." *Journal of Music Theory* 11 (1967): 32–59.

Gauldin, Robert. "The Cycle-7 Complex: Relations of Diatonic Set Theory to the Evolution of Ancient Tonal Systems." *Music Theory Spectrum* 5 (1983): 39–55.

Johnston, Ian. *Measured Tones: The Interplay of Physics and Music*. New York: Adam Hilger, 1989.

Kramer, Jonathan D. *The Time of Music: New Meanings, New Temporalities, New Listening Strategies*. New York: Schirmer Books, 1988.

Krantz, Richard J., and Jack Douthett. "A Measure of the Reasonableness of Equal-Tempered Musical Scales." *Journal of the Acoustical Society of America* 95 (1994): 3642–50.

Krantz, Richard, Jack Douthett, and John Clough. "Maximally Even Sets: A Discovery in Mathematical Music Theory is Found to Apply in Physics." In *Bridges: Mathematical Connections in Art, Music, and Science*. Conference Proceedings, ed. Reza Sarhangi, 193–200. Winfield, Kansas: Central Plain Book Manufacturing, 2000.

Krantz, Richard J., Jack Douthett, and Steven D. Doty. "Maximally Even Sets and the Devil's-Staircase Phase Diagram for the One-Dimensional Ising Antiferromagnet with Arbitrary-Range Interaction." *Journal of Mathematical Physics* 39 (1998): 4675–82.

Levarie, Siegmund, and Ernst Levy. *Tone: A Study in Musical Acoustics*. 2d ed. Kent, Ohio: The Kent State University Press, 1980.

Lindley, Mark. "Pythagorean Intonation." In *The New Grove Dictionary of Music and Musicians*, 2d ed., ed. Stanley Sadie. London: Macmillan Publishers, 2001.

_____. "Temperaments." In *The New Grove Dictionary of Music and Musicians*, 2d ed., ed. Stanley Sadie. London: Macmillan Publishers, 2001.

Loeb, David. "Mathematical Aspects of Music." In *The Music Forum*, vol. 2, ed. William J. Mitchell and Felix Salzer, 110–29. New York: Columbia University Press, 1970.

Maconie, Robin. *The Science of Music*. New York: Oxford University Press, 1997.

Madden, Charles. *Fractals in Music: Introductory Mathematics for Musical Analysis*. Salt Lake City: High Art Press, 1999.

Palisca, Claude V., and Ian D. Bent. "Theory, Theorists." In *The New Grove Dictionary of Music and Musicians*, 2d ed., ed. Stanley Sadie. London: Macmillan Publishers, 2001.

Rahn, Jay. "Constructs for Modality, Ca. 1300–1550." *Canadian Association of University Schools of Music Journal* 8, no. 2 (1978): 5–39.

_____. "Coordination of Interval Sizes in Seven-Tone Collections." *Journal of Music Theory* 35 (1991): 33–60.

_____. "Some Recurrent Features of Scales." *In Theory Only* 2, no. 11–12 (1977): 43–52.

Regener, Eric. "On Allen Forte's Theory of Chords." *Perspectives of New Music* 13, no. 1 (1974): 191–212.

Roederer, Juan G. *The Physics and Psychophysics of Music: An Introduction.* 3d ed. New York: Springer, 1995.

Rothstein, Edward. *Emblems of Mind: The Inner Life of Music and Mathematics.* New York: Times Books, 1995.

Santa, Matthew. "Analysing Post-Tonal Diatonic Music: A Modulo 7 Perspective." *Music Analysis* 19 (2000): 167–201.

Scherzinger, Martin. "The Changing Roles of Acoustics and Mathematics in Nineteenth-Century Music Theory and Their Relation to the Aesthetics of Autonomy." *South African Journal of Musicology* 18 (1998): 17–33.

Sethares, William A. *Tuning, Timbre, Spectrum, Scale.* New York: Springer, 1998.

Stutz, Conley, and Beverly Williams. "Ernst Ising." Obituary. *Physics Today* 52, no. 3 (March 1999): 106–8.

Sundberg, Johan. *The Science of Musical Sounds.* San Diego: Academic Press, 1991.

Winograd, Terry. "An Analysis of the Properties of 'Deep Scales' in a T-Tone System." Unpublished, n.d. Cited in Gamer, Carlton, "Deep Scales and Difference Sets in Equal-Tempered Systems," *American Society of University Composers: Proceedings of the Second Annual Conference* (1967): 113–22; and "Some Combinational Resources of Equal-Tempered Systems," *Journal of Music Theory* 11 (1967): 32–59.

INDEX

A

Aeolian (natural minor) mode, 18–21
Agmon, Eytan, 20, 156–157
Aliquant bisectors, 158
Analogies, maximal evenness, 13–14, 27–28, 38–39
    dinner table, 13–14, 27–28
    stepping-stone, 38–39
Antiferromagnetic, 144, 159
Aristoxenus, 3
Ascending melodic minor scale and collections, 68–73, 99–100
    bisectors and, 99–100
    non-diatonic collections and, 68–73, 99–100
Augmented triads, 18, 39, 118, 121, 130–131, 143

B

Babbitt, Milton, 4, 152
Balzano, Gerald, J., 152, 155
Barber, Samuel, 154
Barbera, C. André, 155–156
Beethoven, Ludwig van, 154
Bell, James F., 156
Bent, Ian D., 155
Bisectors, 97–105, 158
    aliquant, 158
    ascending melodic minor collections and, 99–100
    c *vs.* d distances and, 97–105
    collection testing and, 101–103
    definition of, 97
    generated whole-tone collections and, 103–105
    harmonic minor collections and, 99–100
    introduction to, 97, 158
    octatonic collections and, 100–101
Block, Steven, 150, 157
Boethius, 3
Brinkman, Alexander R., 152
Browne, Richmond, 151, 157

C

G

H

I